THE CULTURE OF GERMAN

ENVIRONMENTALISM

Culture and Society in Germany
General Editors: Eva Kolinsky and David Horrocks

THE CULTURE OF GERMAN ENVIRONMENTALISM

ANXIETIES, VISIONS, REALITIES

Edited by Axel Goodbody

Berghahn Books
New York • Oxford

First published in 2002 by
Berghahn Books
www.berghahnbooks.com

First paperback edition published in 2004

Library of Congress Cataloging-in-Publication Data

The culture of German environmentalism : anxieties, visions, realities / edited by
Axel Goodbody.
p. cm, --(Culture and society in Germany ; v. 5)
Includes bibliographical reference and index.
ISBN 978-1-57181-797-6 (hbk: alk. paper) -- ISBN 978-1-57181-670-2 (pbk:
alk. paper)
1. Environmentalism --Germany. I. Goodbody, Axel, 1950- II, Series.
GE199.G3 C85 2002
363.7'00943--dc21

2002066601

British Library Cataloguing in Publication Data
A catalogue record for this book is available from the British Library.

Printed on acid-free paper

ISBN 978-1-57181-797-6 hardback
ISBN 978-1-57181-670-2 paperback

Table of Contents

List of Tables

List of Figures

Foreword

This book differs from existing political and sociological studies of Environmentalism in Germany in its focus on the culture of the environmental movement. Besides including up-to-date assessments of environmentalism as a social movement and a political force, it is also concerned with explaining its prominence in German political culture. On the one hand, it points to the collective memory and imaginary accumulated in layers of cultural tradition through words, images and events. The responses to modernisation of generations of cultural critics and their conceptions of nature are significant factors impinging on environmental concern. On the other, it asks how contemporaries' concerns have found expression in books and films, and what role these have played and might play in promoting awareness of the need for change.

The book begins with statements by two practitioners, a journalist and a writer. Jürgen Krönig is a London-based political correspondent for the German weekly *Die Zeit*. Environmental questions have featured prominently alongside British politics, globalisation and the media in his many contributions to the press, television and radio in Germany, Switzerland and Britain. His highly critical account of German environmental journalism, stressing its part in contributing through alarmism to Germany's international reputation as a nation of environmental hysterics in the 1970s and 1980s, and the subsequent abandonment of independent critical investigative journalism for lifestyle reports in keeping with the revival of hedonist consumerism in the 1990s, is informed by his own experiences, and draws comparisons between the situation in Germany and Britain.

The Bavarian novelist, essayist and cultural critic Carl Amery has been described as 'the Nestor of the German environmental movement'. He was a leading promoter of environmental awareness throughout the 1970s, and a co-founder of the Green Party. In the interview recorded for this volume, he speaks about the origins of his environmental concern, his politics, the role which the writer can play in raising consciousness and appropriate literary forms and techniques. These 'experiential' contributions raise issues concerning the rise and subsequent decline of the movement, assumptions, hopes and disappointments, and the scope for change, which are developed and analysed in the following chapers.

Chapter 3 is an introduction to the political, social and cultural dimensions of German environmentalism, which sets out the rationale for the volume. This is followed by a section comprising four chapters examining the past, present and future of the environmental movement. Chapter 4 establishes the historical context of environmental concern in Germany, examining links with the internationally influential tradition of German cultural criticism. The next three chapters are devoted to political and sociological aspects. They survey the development of the environmental movement with particular emphasis on the shifting relationship between grass-roots activists and the Green Party, and investigate the present state of environmental concern in Germany, asking whether parties and interest groups have led the public with their agendas or merely followed existing trends, and examining discrepancies between environmental consciousness and eco-friendly behaviour. The concluding chapter in this section (Chapter 7) takes a critical look at the future prospects of the movement as a whole and the Green Party in particular.

The final section of the book is devoted to literary and filmic discourses. It opens with an extract from a recent publication of Carl Amery's, and an extended commentary reviewing the author's practical campaigning, prose fiction, essays and statements on the role of the writer in environmental debate (Chapter 9). Chapter 10 examines public debate on the environment in the GDR, its part in the struggle for political reform in the seventies and eighties, and its cultural transformation in narrative prose and poetry. Continuities and contradictions in the representation of nature and the environment are discussed in Chapter 11 with reference to German films ranging from the Mountain Films of the 1930s to Tom Tykwer's *Wintersleepers* (1997). Finally, Chapter 12 illustrates and reflects on the contribution of East and West German children's books to environmental education.

Three arguments unite these contributions from the disciplines of history, political science, sociology, literary and film studies. The first is that despite the seemingly ultimate achievement of the German Greens in becoming coalition partners with the Social Democrats in Federal Government in 1998, the environmental movement in Germany is in decline. The uniquely successful blend of anxiety about the environment and technological developments with self-realisation and self-expression which characterised the 1970s, a time of relative economic security and idealistic concern with the quality of life, has yielded to 'eco-optimism', youthful hedonism and the pursuit of affluence. As Blühdorn points out, the environment is no longer a rallying cry for social, cultural and political innovation. Though environmental concern remains at a relatively high level in Germany compared with other nations, the future of the Greens as a political force has come to look doubtful.

The second is that the specific characteristics of German environmentalism may be explained not only by economic, political and social circumstances, but also by cultural tradition and developments in contemporary German culture. Environmental debate appears at times to have been driven as much by human anxieties arising from the violation of cultural norms and expectations, as by physical conditions or objectively existing problems. This does not mean that environmental 'crisis' can be dismissed as the 'construction' of a specific society, at a specific juncture, or even of particular social interest groups. The gravity of the challenge in the coming decades is undeniable. However, it opens up a perspective for the understanding of the puzzling differences between national cultures in the importance they attach to individual issues and aspects of the environment.

The third, related argument is that culture not only reflects changes in social values and attitudes towards the environment, but also participates actively in the construction of perceptions of nature and our relationship with it. Through narratives and images, literature and film play a part in determining the way in which environmental problems are framed, publicised and remedied.

My thanks are due to Eva Kolinsky for initial encouragement to undertake a volume in the 'German Culture and Society' series, and to David Horrocks for editorial advice. The DAAD supported a month's research in Germany in 1998, during which I interviewed Carl Amery and collected material for the chapter on his writing. The British Academy, the DAAD and the Department of European Studies and Modern Languages at Bath are thanked for their support for a Colloquium in March 1999, at which drafts of several chapters were presented as papers. The Arts and Humanities Research Board and my own Department funded a year's research leave in 1999/ 2000 during which, inter alia, my contributions to the volume were written, and the initial editorial work was completed. Finally, I thank my colleague Ingolfur Blühdorn, and also Richard Kerridge and other members of the Association for the Study of Literature and Environment (UK), for numerous conversations in which the issues at stake here became clearer to me.

Axel Goodbody

PART I

From the Practitioner's Standpoint: Promoting Environmental Awareness in Twenty-First Century Germany

CHAPTER 1

Eco-Journalism

Jürgen Krönig

In *The Fading of the Greens* (Bramwell 1994) Anna Bramwell predicted the decline of environmental politics in the industrialised countries of the West. The title of her book could serve as a telling metaphor for the plight of Green journalism as well. If and when the Green movement is in crisis, or even in decline, journalism is affected by this development too. The German Greens are presently going through one of the most difficult times since their beginnings in the 1970s. This may seem a strange statement in view of the fact that they have never held more political power in their short history. They are part of the national, Federal Government in Berlin and they have formed various regional coalition governments together with the Social Democrats, not least in the largest German state of Northrhine-Westphalia. Nevertheless they seem in decline. Results in various regional elections as well as opinion polls tell the same story – stagnation or decline. Some observers think that we may even be seeing the beginning of the end of the Greens as a political force. Not of environmentalism, of course, but of a Green Party able to attract the necessary 5 percent of votes, the hurdle a party in Germany has to overcome if it wants to gain parliamentary seats. Recent election results were not merely frustrating for the Greens, especially in the light of the post-Kohl crisis of the Christian Democrats – the trend which emerged must also worry them deeply: the young voters are turning away. The Greens are losing their appeal especially to the young generation. In the eyes of many young Germans they represent a parent-generation that knows everything better, that has no fun at all, an ageing generation of spoilsports, with whom no one wants to have anything in common. It doesn't help that the Green ministers in the Schröder

Government have not proved to be very competent politicians, apart from Joschka Fischer as Foreign Secretary. Take for example the Minister for the Environment, Jürgen Trittin: he made a complete mess of the core element of Green politics, the phasing out of nuclear energy. He started ill-prepared, had to retreat, and now has to live with a compromise of thirty years, which is far removed from the original Green demand of a speedy farewell to nuclear power. All this of course is reflected in the German press coverage. The Greens no longer have a very good press. Their journalistic sympathisers, once so numerous, have shrunk significantly in number. There is disillusionment on the part of Green supporters, and a lot of gloating amongst those who never really thought highly of the Greens and ecological issues anyway. Environmental journalism in Germany reflects the changing attitudes of society and the changing fortunes of the Green Party. All this stands in sharp contrast to the situation in the 1970s and early 1980s, when the Green Party emerged triumphantly from obscurity.

The First Phase

In the first phase, the environmental movement occupied the moral high ground, and environmental journalism was its important messenger. There was a lot of idealism and conviction around, combined occasionally with a tendency to be self-righteous – only ignorant, greedy forces in society could argue against the frightening extent of environmental degradation, could ignore the escalating destruction and pollution. Global threats like ozone depletion, the greenhouse effect and the destruction of rainforest or temperate forests were the big subjects constantly talked and written about. The peak in environmental concern was reached when even a politician like Margaret Thatcher, influenced by Britain's United Nations Ambassador, Sir Crispin Tickell, delivered a speech in which she postulated that 'mankind had unwittingly and unwillingly endangered the planet'. Al Gore, Vice-President of the United States, published a book on global Green philosophy in the 1990s, and titles like *The End of Nature* (McKibben 1990) made it onto the bestseller lists in most western countries, including Germany. All this indicated how deeply Green ideas had penetrated even sections of the political and economic classes.

This was the time when in the United States mainstream magazines like *Time Magazine* and *Newsweek* published title stories about impending global disasters and the *Spectator's* title page showed London landmarks like Big Ben rising out of the sea. *Time Magazine's* personality of the year was Gaia, our planet Earth, whose life support systems were being eroded and destroyed. Even a magazine like *Business Week* presented to its readers in the financial industries and stock markets a title story about global warming and rising sea levels. One should not forget the enormous impact of the Chernobyl disaster. The near meltdown of the nuclear reactor in the Ukraine caused deep concern all over Europe. Nowhere more, however, than in Germany itself. The intense environmental concern expressed all over the western world strengthened environmental journalism in Germany. Media institutions and journalists who so far had been immune against Green issues reacted to this development. Suddenly it became fashionable to write about issues one would not have touched a few years earlier. Leading lights in the German media scene usually take some of their inspiration from observing the most important, prestigious foreign publications, especially those in the United States and Britain. At the end of the eighties, when global conferences about climate change and ozone depletion mushroomed, environmental journalism in Germany was more *en vogue* than ever before. Even some of the bigshots of German journalism were forced to show some interest – editors or leader article writers of papers like the *Frankfurter Allgemeine, Süddeutsche Zeitung, Die Welt* and *Die Zeit*, who preferred to concentrate on the classic grand topics like foreign politics, the strategic balance, star wars and disarmament – the stuff of thousands of leader articles – decided that they could not abstain from ecological topics any longer. Green topics had until then always been left to younger writers, because they were regarded as a bit soft, if not dubious or cranky. Not any longer.

An amusing example of this sea change concerns my own paper, *Die Zeit*: at the end of 1990 a leader article was published, written by the editor Theo Sommer, a well-known figure on both sides of the Atlantic. In his article he demanded urgent global action to combat the greenhouse effect and ozone depletion – time was running out, his main message ran, governments needed to act. A few months later a young journalist congratulated him on this article. The author seemed slightly embarrassed – he assured his young colleague that it was not really that urgent, that there was plenty of time left to sort things out.

German journalists of the older generation were never part and parcel of the younger, environmentally sensitive scene; they had never shared their hopes and sympathies. Even new scientific evidence and greater political willingness to take action against global environmental threats did not cause a real change of heart in the upper echelons of German journalism; but the overall climate forced even them to pay at least some lipservice to Green issues.

This was the time when Germany's reputation as being environmentally oversensitive, if not hysterical, was born abroad. *Rich, Bothered and Divided* was the telling title of a book about modern Germany, written by David Marsh, an astute observer of my country shortly before unification and the collapse of Communism (Marsh 1989). Even the infamous Chequers conference on German national characteristics held by Margaret Thatcher explicitly minuted the German tendency to be permanently worried, especially about all sorts of environmental dangers and risks. It was environmental journalism which helped create and shape this collective state of ecological concern in Germany.

However, it was not to last. The 1990s, especially the second half of that decade, saw a significant shift in emphasis. Various factors contributed to the decline in the importance of environmental journalism. There is the boredom factor, which should, as far as journalists and their audience are concerned, never be underestimated. A number of journalists found it just boring to swim with the tide and repeat the frequently advanced warnings; they started writing articles which played down if not ridiculed global ecological concerns. Some of them did so purely to be controversial. A well-known reporter from *Der Spiegel* admitted quite openly that he had decided to write against the potential danger of the greenhouse effect because he wanted to be different. He was tired of being just one of many writers warning about global warming or the rapid destruction of rainforests. It is indeed a fact of life in journalism that it often pays to be controversial or to take a different stance. Your voice gets heard more clearly, regardless whether what you say or write is true. However, boredom and cynicism are not the only reasons for the waning of environmental journalism in Germany.

The Green movement and its journalistic sympathisers had to pay the price for too much doom and gloom. There had been an overkill of alarming articles and reports during the first phase of what I call the environmental age. Alarmism was widespread. In Germany the groundwork had been laid by authors like Erhard Eppler, a Social Democrat, who wrote *Ende oder Wende?* (The End or a Turning Point? – Eppler 1975). Another highly influential writer, coming from the other side of the political spectrum, was Herbert Gruhl, until 1974 a Christian Democrat MP, who shattered West Germany's 'Wohlstandsgesellschaft' (affluent society) with his chilling bestseller *Ein Planet wird geplündert. Die Schreckensbilanz unserer Politik* (The Pilfering of the Planet. The Terrible Consequences of Today's Politics – Gruhl 1975). A few years later the mail order publishing house 'Zweitausendeins' sold nearly a million copies of the 'Global 2000 Report to the President' commissioned by Jimmy Carter (Kaiser 1981). It was a voluminous book of more than a thousand pages, in which every aspect of environmental concern was covered – from overpopulation and shrinking resources to global warming – written by the world's leading scientists. Books like these, and there were many more, influenced and helped to shape environmental journalism in Germany. Radio and TV stations appointed environmental editors (Redakteure), more air time and space was given to Green topics; *Der Spiegel* alarmed the informed public with a series of dark, threatening title stories, about every possible threat to mankind and nature, not least 'Waldsterben' (forest dieback), which triggered special anxiety in the collective consciousness of the Germans. New environmental publications appeared, for instance the magazine *Natur* – not to be confused with the British science magazine. There was also the enormous impact of the report of the Club of Rome (Meadows 1972) with its very precise, and (as we now know) false predictions about the time when resources would be finished if exploitation was to continue at the same pace – oil in 1992, gas in 1993.

This permanent doom and gloom, the prophecies about impending ecological disasters, have in the long run undermined the position of the Green movement and the credibility of environmental journalism. People became wary of what was called the 'Green religion'. The increased scepticism helped to fuel an anti-Green backlash which was beginning to have an impact. Big multinational companies, the coal and oil industry, not least the battle-hardened chemical industry, made a concerted effort to roll back the environmental movement. They started PR campaigns, they targeted the political and scientific elites successfully. Ecological arguments were turned on their heads, scientific studies financed by institutes set up by big corporations tried to prove that global warming was nothing but a myth. American thinktanks produced a flood of papers which targeted green 'legends' and scientists published books with telling titles like *Small is Stupid* (Beckerman 1995) or *Life on a Modern Planet* (North 1995), refuting ecological myths.

Germany Today

In the last decade of the twentieth century environmental issues started playing a much less important role in public consciousness. The priorities of the population changed considerably. It is a well-known phenomenon that if the economic cycle produces a downturn or a recession, the interest in ecological issues recedes. Economic recession causes environmental depression. The change in question went beyond a mere reaction to less favourable economic circumstances. Nevertheless, despite the priority of economic and social issues and the deep worries about unemployment, Germany is not just returning to the age of pre-environmental innocence or ignorance. Environmental issues have retreated, the media have put them on the back burner, but ecological concerns have not vanished.

Important elements of the Green message have more or less been accepted by the majority, while a significant minority even acts accordingly, tries to use less of the finite resources, to save energy and water, and separates rubbish for recycling. (Even this environmentally aware minority cannot be absolutely sure it is doing the right thing – a report on German Television in 1999 emphasised the negative consequences of lower water use: the pipes are not getting flushed out thoroughly enough any longer.) However, the majority of people, even though they accept the reality of long-term risks for the planet, no longer regard environmental problems as a reason for politicians and governments to act as urgently as possible. Here, it seems, mainstream journalism agrees with its audience: the big stories have all been told, the important environmental problems have all been recognised and analysed. One cannot go on doing this again and again. What is more – the exaggerated prophecies of impending disaster have not materialised. There is also an element of resignation – one can't help it anyhow; the march of progress in combination with the growth of population worldwide can't be stopped. Some of the global ecological damage done is irreversible anyhow, however great it may be.

Added to this are other, even more powerful influences, which shape public and published opinion in Germany – hedonism, materialism and consumerism. Sometimes it is driven by an attitude of 'Nach mir die Sintflut' (*après moi le déluge*). The majority look at the world with a strong optimism, guided by the belief that technological progress will sort out all eco-problems in the long term anyhow, an attitude to be found especially among younger Germans.

For environmental journalism this mix of feelings, emotions and trends has consequences. I have heard from colleagues in a number of papers that it has become much more difficult to find an open ear for Green issues in editorial conferences. Rarely are ecological topics lifted onto the front pages or displayed prominently inside the papers. Lifestyle journalism has replaced Green journalism to quite an extent. 'Jammerartikel' (lamentations) are not popular with editors, because they create 'bad vibes' among the readers and – a very important point in times of falling circulations in Germany's print media – they are not popular with the advertisers, as I have been told by the head of marketing of a famous German paper.

The readyness of journalists to fight for uncomfortable environmental topics has grown weaker, not only because of the change in the public mood, but also because of another factor. There is a feeling of regret for one's own sins of alarmism. A point in case was the *Brent Spar* incident. The dispute over the planned disposal of the Shell oil platform by dumping it into the North Atlantic triggered a near-hysterical and almost exclusively one-dimensional approach in Germany. Greenpeace played a dubious role in this media drama, feeding false information about the toxic content of the *Brent Spar* to journalists. The facts were not checked, but used to feed the emotions of anger.

Of course alarmism in itself must not necessarily and always be wrong. Things may sometimes be every bit as bad as the alarmist's warning makes them out to be. Anyhow, alarmism may have an important function: it can help raise awareness of problems which undeniably exist. Alarmism can increase sensitivity. However, there is always the danger of creating hysteria, which in turn will lead to the equivalent of cold turkey, disillusionment and cynicism. All too often alarmism has proved to be counterproductive. It has allowed the other side, vested economic interests, to brand everything the alarmists said, including justified and rational warnings, as irrational.

In contemporary Germany environmental journalism does not play as important a role as in the 1980s or before. The overall character of ecological journalism is today neither alarmist nor indifferent, it is somewhere in between. The dominance of the electronic media has changed the character of reporting on environmental topics. Television is picture-driven. It tends to concentrate on disasters and accidents. The print media follow. The coverage of ecological topics today is more event-driven than before; as a consequence one rarely finds an analytical approach or the attempt to put accidents or events into a wider context. Environmental journalism is in this respect afflicted by a general trend which can be observed in the wider media industry. The process of 'dumbing down', which undeniably exists, leaves less space for serious environmental concern in the electronic and print media. There has been a clear shift to a kind of reporting which concentrates on and in most cases vastly exaggerates the dangers for the personal health or life of the viewer/reader. Bovine spongiform encephalopathy (BSE) was a prime example.

On top of this, the younger generation of journalists does not share the concerns of their peers. Quite often they are highly critical of traditional environmental journalism. They are fascinated by the rapid progress of modern information technology and draw optimism and belief in progress from this. This tendency is enhanced by an urban or metropolitan lifestyle, quite often far removed from any direct knowledge or experience of the natural environment.

There is another tendency which might become more important in future. Most media outlets have a few journalists who specialise in environmental issues; they are not necessarily part of the science section, which in many cases wants to have the overall say about so-called Green issues, and has managed to achieve this. As a consequence, an interesting gap has opened between environmental and science journalism. The latter is quite often much less worried about potential risks of new technologies and tends to play down or ignore their potential negative impact on the environment. One example is the ongoing dispute on intensive farming methods and the health risks of pesticides and herbicides, another the conflict over the introduction of genetically modified organisms. Science writers tend to believe in the ability of science to deliver progress without negative consequences. They tend to follow the arguments of their scientific peers.

However, science has changed. To quite a degree it has turned into 'corporate science'. Scientific institutions are more than ever before dependent on research grants from business and State institutions. The field of research into genetically modified organisms (GMO research) for example is funded to quite an extent by biotech multinationals like Monsanto. There are few independent scientists left, a fact rarely admitted by ministers. Science journalists, who often have a background in science themselves, wish to be accepted as equals by 'real' scientists. They accept their results more readily and are less willing to question research results whose interpretation may be coloured by the vested interests which financed the research in the first place. The *New Scientist* referred in a recent self-critical article to the 'dominance' of corporate science serving the interests of multinationals. Universities too are losing their independence more and more, because to a growing extent they have to rely on funding from big corporations. Even if governments are really looking for independent expert advice, which is not always the case, they have enormous difficulty finding independent scientists who are not directly or indirectly dependent on the same companies, whose new products, be it GMOs or pesticides, they have to evaluate. One would expect journalists to be aware of this fundamental problem and to act accordingly. However, German science journalists, like most of their colleagues in Britain, rarely live up to this expectation. To illustrate the point, *Der Spiegel* is well known for its tendency to dramatise or occasionally go completely over the top.

Yet not once during the last decade have the science pages of this journal contained an article looking at the possible role of organophosphates (OPs) in various neurological disorders and diseases, like Gulf War syndrome, BSE, or the poisoning of nearly 1000 farmers and farmworkers in Britain after dipping sheep in a cocktail of toxic chemicals. Organophosphates are pesticides based on nerve gas. They have become one of the most successful products of agrobusiness. Against all the odds, there is some hard scientific evidence available indicating that OPs are one factor responsible for the outbreak of BSE in Britain. Despite various research papers published in peer-reviewed journals, coming for instance from the University of Cambridge, no British or German science journalist has taken this up or even reported on it. They have preferred to play it safe, following the guidance of official scientific institutions and government officials. There has been a disappointing lack of cool, detached journalism, trying to investigate glaringly obvious contradictions and shortcomings of the official BSE and variant Creutzfeldt-Jakob disease (VCJD) position. Instead the years of the BSE crisis revealed another face of modern journalism: a more short-term, sensationalist approach is affecting the output of all media outlets. There is less room for balanced journalism, for analysis, for seeing and explaining events in a wider context or observing processes instead of going for a crass headline or extraordinary story. The trend has touched even the best of the so-called quality papers and magazines. Environmental stories have changed too, of course. If they make it into print at all, they tend to be sensationalised and hyped, only to miss the real point. Television, the most powerful medium of our age, is driving this process forward. It is the most effective instrument in creating a culture based on consumption and commerce. Its whole existence, one could argue, relies on ever-increasing consumption. The message to be happy and consume does not go well together with warnings about shrinking resources, water scarcity or the dramatic loss of topsoil. Nor are editors over-fond of such warnings who think, or even know, that all too critical articles about oil and life science companies, or too much bad 'mood music', could have an impact on the volume of advertising their paper depends on. As far as documentaries on TV are concerned – they have to be racier and sexier to survive on the main channels – even public broadcasters, drawn into an ever more intense ratings war with their commercial rivals, are giving in to infotainment. This is not a trend peculiar to Germany, far from it. It is more pronounced in Britain or the United States. But it is fast gaining ground around a world in which globalisation and the information revolution have led to a dramatic increase in the volume, intensity and speed of communication and cultural exchange. At the beginning of the new millenium we are confronted with a strange contradictory situation, in the industrialised world in general, and of course in Germany. Germany is an environmentally more advanced country than most of its neighbours, or so it likes to believe. At the same time the Germans are perhaps even more than others in the grip of material values, dancing around the golden calf of consumerism and luxury consumption. Yet the Green beast is not dead, it is only slumbering. When the global problems become more obvious and action is needed more urgently, it might awake abruptly. Journalists, having known it all along of course, will be its eager supporters.

CHAPTER 2

Lianas Across the Jungle:
An Interview with Carl Amery

Axel Goodbody

AG Carl Amery, your name is not only associated with novels and
critical essays on contemporary culture and politics, you are
also a veteran environmental campaigner and a Green political theorist. For a generation you have played a prominent role
in both the cultural scene and debate on the environment, first
and foremost in your native Bavaria, but also in Germany as
a whole. You were a member of the influential writers' group
Gruppe 47 in the fifties and sixties, and have since served terms
as President of the *Verband deutscher Schriftsteller* (Association of
German Writers in the Printing and Paper Union), and as President of the West German PEN. You are the recipient of
literary prizes ranging from the Ludwig Thoma medal for
contributions to Bavarian culture to the Kurd Laßwitz prize for
science fiction, and non-literary distinctions including the
Order of Merit of the Federal Republic of Germany, First
Class. One of your most recent awards, the Wilhelm Hoegner
prize of the Bavarian Social Democratic Party, was presented
to you as an 'angular Bavarian Free State patriot, critical
Catholic and environmental thinker'. Your reputation, in short,
is that of a popular writer and controversial essayist, but
equally of a persistent champion of the environmental cause.
So my first question is: When and how did your concern for
the environment begin?

CA In a way I have come *back* to the environment. My family
background is one of conservative South German Catholicism,
conservative in the original sense of the word. As an educated
man and a historian, my father, who took part in the liturgical movement, reflected environmental concerns already in

his writings. The term 'environment' wasn't used in those days, of course, but he was a typical 'lamenter' of disappearing land-scapes. He was annoyed by electricity pylons in the countryside and by factory chimneys in the wrong places. I didn't follow him in this – you distance yourself from things like that growing up. But I realised when I came to think about environmental problems that much of my emotional involve-ment came from my background. It is a powerful current in German cultural tradition, as indeed in English culture too. The Romantics were profoundly sceptical about the progress of civilisation. Tieck, for instance, and Wordsworth. This has played a central role in German thought in the last two hundred years, and not only in the positive sense.

But when I began to write, it was not about such things. Instead I wrote a critical little book about Catholicism in Germany, *Die Kapitulation oder Deutscher Katholizismus heute* (Capitulation or German Catholicism Today: Amery 1963). I called it 'real existing' Catholicism, by analogy with socialism in the GDR. It caused a tremendous stir at the time, though if you read it now it's rather harmless. Any priest could write it today without getting into trouble. But back in 1963 it was a sensation, because of the suppression of the past by the Catholic church in post-Nazi Germany. I was suddenly in great demand as a public speaker – and was accused of not knowing enough about modern theology. So I decided to try to catch up. At the time the *Paulus-Gesellschaft*, which was founded by a doctor, was working towards a reconciliation of Marxism with Christianity, and organised some important conferences. It captured the spirit of the times. A book published by an American (Cox 1965) argued that Christianity itself was ulti-mately responsible for the secularisation of the western world, the humanising of the earth, and the abolition of taboos arising out of the *tremendum* of nature. I had reached the point of recognising the dangers in this. I included a parodistic montage of extracts from the minutes of a meeting between theologians and (mainly Czech and Polish) Marxists in my book *Das Ende der Vorsehung* (The End of Providence: in Amery 1985a: 172–6). It was called 'Moby Dick in Marienbad'…

AG The passage about the whale threatened with extinction, which goes to the conference for help?

CA Yes, the piece about the whale! That was something quite new. As it happened, the radical student movement in America was just turning its attention to environmental issues at this time.

I was familiar with the movement – my wife is American, and I've written for American magazines – and was about five years ahead of the general public in Germany. So there were two critical masses – I usually need two critical masses to come together in my books. That is how I became involved in the environmental movement. I spoke on environmental issues as early as 1970 in election campaigns, at a time when hardly anyone else was addressing them. *Das Ende der Vorsehung* came out in 1972. Since then I have come back to the subject again and again. It gave me a second wind. Suddenly I had a perspective again, a feeling I was needed in the cultural scene, and had something to say. Another consequence was that I gained a completely new circle of friends. I had to get used to scientific thinking, not scientific knowledge – it was too late for me to start pursuing that – but to begin familiarising myself with scientific perspectives, the scientific orientation. They were very interesting people, and I get on better with them now than with many literary people! A third consequence was that neither the strident criticism of the old Marxists nor their more recent capitulation have unduly affected me. I had contacts for instance with the Bulgarian Academy of Sciences, which suddenly discovered ecology, and gave interviews in Moscow for *Voprosy filosofii*. I found I managed quite well with my own concepts and categories, despite having only a super-ficial knowledge of Marxism. The collapse of the West German Left in the eighties did not affect me either.

AG Around 1970 you campaigned for the Social Democrats, and later for the Greens.

CA I was a founder member of the Greens in the late 1970s, a sort of *spiritus rector*. My book *Natur als Politik* (Nature as Politics: Amery 1976) served as a *vademecum* of the Greens.

AG Are you still involved in the Green Party?

CA Yes, this evening for instance I'm going out to Grafing to give an election support speech. I'm not physically able to do so much as I used to. And I don't see myself as a Party Horse. I work mainly above party politics. But if you want to give me a party label I would still call myself a Green.

AG You have also played an active role in local Citizens' Initiatives and NGOs. I believe you were a member of the *Gruppe Ökolo-gie*, which brought together leading conservationists with some of the first German environmentalists in the early seventies.

CA Yes, that was the first organisation. It was a group of concerned scientists – the first 'concerned scientists' were people like Paul Ehrlich and Gordon Rattray Taylor in the United States, and to a certain extent Rachel Carson. But Rachel Carson was more than just a concerned scientist. Her book, *Silent Spring*, was the basic text for a whole political movement. Like Aldo Leopold's cult book later on.

AG Was *Sand County Almanac* important for you personally?

CA No, I only read Carson (and Leopold, who I actually regard more highly) much later. I have them there on my shelf, they're a source of useful quotations! But they didn't influence my development.

AG And you were a founding member of the Schumacher Society?

CA Yes, the Schumacher Society here in Munich, which has been quite influential though it hasn't many members. They started the excellent journal *Politische Ökologie* (Political Ecology). There is a flourishing branch in Stuttgart too. It has nothing to do with celebrating E.F. Schumacher's life and work. We chose the name at the time because we wanted something like the Party Cultural Foundations. Schumacher, who I knew, and still regard very highly, seemed to us at the time the ideal choice, as an environmentalist pioneer of international stature who was born in Germany.

AG Did you also take part in the Peace Movement? Were you at the Mutlangen blockade?

CA No, I wasn't. For personal reasons. As Günter de Bruyn puts it so aptly in his memoirs (de Bruyn 1992), sometimes I stand outside myself, and find myself in danger of becoming a bit of a buffoon. It's not my scene. I've given lots of speeches, at demonstrations in Wackersdorf, Landshut and Klein Ohu, and been very happy to be involved, but I couldn't lead myself to take part in such solemn acts of public disobedience.

AG Carl Friedrich von Weizsäcker has written of a 'shift of consciousness' being essential if the human race is to survive. Various social institutions are involved in the promotion of ecological consciousness – the state, the political parties, schools, the media and the culture industry. You have spoken of the necessity of giving the ecological perspective concrete expression in art, literature and the study of literature – also in theology. What particular role can literature and fiction play in this process?

CA That is the thousand dollar question! I don't think the perspective we really have to internalise can be promoted effectively by means of literature. I've written about this in some of the essays in *Bileams Esel* (Bileam's Ass: Amery 1991). I'm not sure whether the confidence I had at the time is really justified, I see considerable difficulties here. You can discount mainstream literary publishing anyway, which is primarily concerned with stories of hatred, love and betrayal. Take, for example, the books discussed in cultural programmes on television like the 'Literary Quartet'! In general, you can say the role of literature in society we've become used to is, to use a theatrical image, one of inter-social movements taking place before a background or backdrop in which nature only plays a modest part. It would be interesting to develop a systematic approach similar to Marxist literary theorists – Lukacs, Benjamin and Brecht too, to a certain extent – re-reading older texts from an ecological point of view. Homer and Virgil offer wonderful possibilities: *Majoresque cadunt altis de montibus umbrae.* If I weren't so old myself, I might have got involved. But perhaps it's better, maybe my talents lie elsewhere. In my view one of the first things to do would be to devise a literary theory based on an ecological perspective.

AG Some studies of American literature have recently been written from an ecocritical perspective. But the term 'ecocriticism' sounds ugly in German, and academic trends and fashions are different here.

CA At the moment the outlook isn't promising here. But the situation in literary criticism is no different from the general ecological situation in Germany: there's not much going on!

AG Of course the United States possesses a strong tradition of nature writing going back to Thoreau, to which Aldo Leopold and Wendell Berry belong. This genre is less prominent in Germany.

CA I would add William Carlos Williams, and include younger writers like Snyder. Yes, there's something in that. But Hitler is to blame of course. He prevented us from reflecting on such things in Germany for decades.

AG Yes, the ideological distortion of Green arguments in the Third Reich may ultimately be responsible for the suspicious attitude towards environmentalism prevalent in the humanities in

Germany. And I would like to return to your book on Hitler
later. But may I first ask you about your own experience? You
have a reputation both as a 'green philosopher' and as a writer.
What response has your theoretical writing had? Did it provoke
public debate? Have people written to you, for instance?

CA Oh yes, I played a part in the formulation of the Green
programme. *Natur als Politik* came out in 1976. It was of greater
practical importance at the time than *Das Ende der Vorsehung*,
though more thought had actually gone into the first book.
Natur als Politik sold especially well in paperback, whereas sales
of *Das Ende der Vorsehung* had been mainly in hard cover. Tens
of thousands of paperback copies of the second book were sold.
That shows who it was read by – young people who hadn't so
much money but wanted to be informed. I had quite a lot of
feedback from readers, and, to the extent you can ever be satisfied with yourself, I can't complain of the public response.

Every now and again I still write something for the press, and
people comment on it. But they don't always agree with me.
For instance I crossed swords with Hans Küng recently. A series
of articles was published in *Die Zeit* about an initiative of
Helmut Schmidt's, the former Chancellor. Schmidt brought
together ex-prime ministers from five continents in an *Inter
Action Council*. They wanted to formulate a codex of fundamental human duties and submit it to the UN. I immediately
recognised the hand of Hans Küng, president of the Global
Ethic Foundation in Tübingen. In the course of the debate,
which was quite heated – the liberals spoke of a suppression of
human rights (what does duties mean anyway, we don't have
any, that was their standpoint) – Küng outed himself as the
author of the catalogue. There was very little of ecological relevance in it. He explained it was based on four commandments
supposedly common to religious ethics all over the world: thou
shalt not lie, thou shalt not steal, thou shalt not kill, thou shalt
not fornicate. It was a sort of basic catalogue, which could only
be interpreted metaphorically as having any ecological relevance – do not steal from your grandchildren, the killing of
nature, etc. I published an admittedly rather cheeky contribution in *Die Zeit*, called 'Ptolemäer und Plattweltler' (Ptolmaeans
and Flat Earthers: Amery 1997). I suggested Küng was a Ptol-
maean, that is, knowing the earth is round, but still placing man
at the centre of things. An anthropocentric, as opposed to the
Copernican environmentalists. The flat earthers were those still
operating on the two-dimensional principles of Adam Smith's

utilitarian materialism. That is how I saw the debate. Küng was not amused. It's crazy – he was given millions by Coca Cola, and can do what he likes now without a worry. It's not like that in the ecological movement. But I can't complain of the response to my writing on the whole.

AG And how about your novels?

CA My prose is sparse, and I make considerable demands on my readers, so my print runs have not broken any records. From the critics my novels have drawn an amusing response. In a typically German way, they asked: 'Is he serious, or just a writer of popular fiction?'! It was completely unimportant to me. I've been called the doyen of German Science Fiction, and I am still practically revered for *Der Untergang der Stadt Passau* (The Fall of the City of Passau: Amery 1975), and of course *Das Königsprojekt* (The King Project: Amery 1974). *An den Feuern der Leyermark* (The Fires of the Leyermark: Amery 1979) too. The latter is one of the few German alternative time stream novels, asking what would have happened if history had been different. I still think highly of it. It had a schizophrenic reception: On the one hand there were the establishment mainstream critics, who didn't understand the book. Ironically, they cost me readers by saying I was too esoteric. On the other hand, I satisfied quite a respectable readership in the science fiction community, people looking for good SF, time machine problems and so on. These are things Marcel Reich Ranicki doesn't know the first thing about! I enjoy writing anachronistically, breaking through time barriers. This is the structure of *Die Wallfahrer* (The Pilgrims: Amery 1986) too, probably my most important novel. It was on my part from the outset an attempt, at least unconsciously, to develop a literary form appropriate to the perspective essential for the survival of humankind. It is for others to say whether I have been successful or not.

AG Would it be right to suggest you first treat your themes discursively, and then go on to clothe them in literary fiction afterwards?

CA Not at all, never! I can say that quite definitely. Fictional ideas come of their own. As the horror writer Stephen King has said: the boys in the basement have to do their shovelling first. There are things an author can't plan in advance. Especially in the fictional texts, I have always needed two critical masses, which converge to yield the material for a book.

AG Let's see: in the seventies, you first wrote the theoretical book *Das Ende der Vorsehung* (1972), followed by the story *Der Untergang der Stadt Passau* (1975), then a theoretical book again, *Natur als Politik* (1976), followed by *An den Feuern der Leyermark* (1979).

CA I also wrote *Das Königsprojekt* in 1974. It was really my first novel. And I had already written two novels twenty years before, *Der Wettbewerb* (The Competition: Amery 1954) and *Die Große Deutsche Tour* (The Great German Tour: Amery 1958), which was pure satire, a bit before its time. But I found my own style in *Das Königsprojekt*. *Der Untergang der Stadt Passau* was really only a finger exercise, even if it became my greatest success.

AG It's a good read, and particularly accessible to young people, of course! But by no means a simple book, for instance the narrative structure…

CA Of course, and I carried it around in my head for some time before writing it down. But I'd say I arrive at the subject matter of my books in much the same way other novelists do. There was no direct dependence on my theoretical books, even for the motif of post-catastrophe depopulation, which I've used several times. I don't think it was directly linked with my non-fiction writing, even if Joseph Kiermeier-Debre suggests in his book that I first had to work through the various ideas theoretically, before I was free to work them up fictionally (Kiermeier-Debre 1996: 199. See also, however, p. 147, where *Der Untergang der Stadt Passau* is cited as an exception).

AG You have of course used intermediate forms between fiction and non-fiction, satirical inserts and documentary passages, in the novels for instance.

CA *Pseudo*-documentary passages! I think that is just a characteristic of modern literature. The nineteenth-century fixation on narrative stream is more or less a thing of the past. Decades ago, long before I knew what I was going to write, I read and learned a lot from Dos Passos, it's all there.

AG Your grim prophecies, warning of seemingly inevitable catastrophe, of the self-extermination of mankind through the exhaustion of resources, pollution of the environment and population growth, have led to your being called a Cassandra. Yet your writing is also decidedly humorous. The reading public has become tired of the environmental alarmism of the seventies, and of apocalyptic scenarios in literature, and now tends to shrug them off. As you yourself have said, the situation is hopeless, but therefore not serious. Has your standpoint changed over time? Do you write differently now from in the seventies?

CA That is hard to say. I think my environmental commitment, leaving my fiction writing aside, has become more concrete. Perhaps my programme was too esoteric to begin with. *Hitler als Vorläufer* (Hitler as a Predecessor: Amery 1998) was to have been written before *Botschaft des Jahrtausends* (The Message of the Millennium: Amery 1994), which, at least in the wind-up chapters, follows on from it. In that sense I've become more concrete and more practical. In my environmental activities too. My work on energy and my part in the struggle against the monopoly of the large power companies is one aspect of a life-long striving for freedom. Since Hitler, who humiliated everyone so terribly – me too, of course – I don't like being taken for a ride, so I try to be a pain in the neck! My motivation is civic, it's a matter of citizens' rights. This is the area I'm most at home in. The district of Schönau in the Black Forest is trying to opt out of nuclear energy. We have collected two million Marks for alternative power generation. And in East Bavaria I have a friend with a small firm. I support him, I put him in touch with people. Quite extraordinary energy savings are possible! Saving electricity is one starting point then. Another is charitable foundations. I am trying to get things going here together with friends. This is a field in which Germany lags behind other countries, and has always lagged behind – well actually, not always. The German Empire and the Weimar Republic had a flourishing system of foundations, ninety-five percent of them Jewish! Through the rape of their funds and the abuse of trust, the idea of foundations has wasted away in Germany. Here again, there is scope for action in the sense of a civil society. In such things I see, if not bridges, then footbridges, lianas across the jungle.

So much for my practical activities. Regarding humour, though, a view of the world determined by true humour seems to me more comprehensive than a tragic one. Humour is, in all seriousness, the most important contribution of Christian civilisation to western cultural tradition. Of course there was Aristophanes' wonderful humour in the ancient world, but the great humorous works were mostly written at the beginning of the modern European national literatures. In Germany there was Grimmelshausen's *Simplicius Simplicissimus*, and in Spain Cervantes. Humour doesn't skate over serious problems, it actually embraces tragedy. Rabelais isn't silent about the murder and violence in sixteenth-century France, and Moliere's *Misanthrope* is one of the saddest plays ever written. I am confident this permeating of the cosmos of experience, feelings and thought with the juices of humour – humour originally meant juice! – brings benefits which tragedy can't.

AG Do you see any contradiction between the seriousness of your environmental message and humour in your novels? If we take the concept 'postmodern', your novels come close to it for instance in breaking down the barriers between high and popular culture, and in your ironic detachment. Perhaps there is even a degree of conscious ambiguity. But at the same time you write in the tradition of the Enlightenment.

CA Yes, leaving aside the literary techniques you have described, I would see myself as belonging to an old-fashioned category for which there is really no German word, the French *moralistes*. In German 'moralist' suggests finger-wagging. Of course that's not what I mean. The *moraliste* is someone who observes the *moeurs* of society, analysing them, exposing their strengths and weaknesses. That is done in fiction as well as discursively. It is the tradition of Voltaire.

AG Have you literary models?

CA In the war I was a POW in America. I read a great deal. Later, after the war, I discovered the British tradition, or rather what was read at the time: Graham Greene and Evelyn Waugh, whom I admired a lot for a time, though less so now. He is a bit journalistic. But I read his great, really biting satires, such as *Scoop* and *Put Out More Flags*, one after the other. Then I was influenced formally by an American professional writer, J.P. Marquand, and wrote my first satirical novel in his style. He had published a pseudo-documentary epistolary novel about Boston Brahmins, *The Late George Apley*. So I got to know these lighter forms in American and English literature.

AG What about Chesterton?

CA Chesterton has been important to me since my youth. As a teenager I devoured his novels. I have just been asked to review the new Chesterton book *Ketzer* (The Heretics: Chesterton 1998), published by Hans Magnus Enzensberger in *Die andere*

Bibliothek (The Other Library). But equally, when I read from *Das Geheimnis der Krypta* (The Secret of the Crypt: Amery 1990) in Prague in 1991 – I've never had such an intelligent public! – it was suggested my affinity with Czech humour would be worth investigating. I was taken aback, but there's something in it. A Schwejk-like, vulgar, vital element, which is perhaps missing in English literature. Maybe it's regionally genetic in origin.

AG Following the decline of environmental activism in the late 1970s, and the shift of political activity towards parliamentary representation, German writing on environmental issues became more ecocentric and irrational. I would like to ask you about the trend towards mythologising nature and resacralising it. *Die Wallfahrer* ends with two alternative scenarios for the end of the world. The first is more or less orthodox Christian doctrine, the Last Judgement. The second, 'heretical' scenario is set in a distant future in which life goes on after the human race has wiped itself out. New species have evolved. Here you bring in the figure of Gaia, the great Earth Mother. Do you think the Earth Goddess can fulfil a function by contributing to the development of a new respect, reverence and humility in the face of nature?

CA I was taking up ideas that were in the air at the time. I participated in a Gaia conference – it was really the first – in Cornwall, with James Lovelock and the American biologist Lyn Margulis. And I was very impressed. Not by Gaia as a *myth*, but as a scheme for explaining the world, the world of living things. I discovered such things were being discussed at church conventions. The Catholics have cleverly retained a dimension of this kind in the figure of the Virgin Mary. She is an interestingly ambivalent figure, and is increasingly regarded as *mediatrix omnium gratiarum*, the mother of god as mediator of grace, which is not so far away from Gaia. The act of creation, the truly sensational process of the emergence of life out of the primeval mud, is logically a female act. The idea of mother nature is practically an anthropological constant. I myself don't believe in Gaia as a religious possibility, that is in the revival of a religion centred on the Gaia principle. But as a semantic figure she has a lot to offer.

AG The ecofeminist movement has a spiritual wing seeking to reinstate goddess worship. But this is not something you would support?

CA No indeed. I regard it with some regret, for the women's sake. I see in it a parochialisation of both movements, feminism and ecology. There can be no question that women have suffered injustice throughout world history. Of course I recognise that.

On another level, feminism has much to offer environmental thinking. I came across an interpretation of the story of the Flood in a highly intelligent journal called *Orientierung,* published in Zurich. It was written by an Old Testament scholar, a man, who specifically described his interpretation as feminist. He told the story of Noah like this: God is angry with humankind and says they must be done away with, then he relents and gives Noah a chance – and with Noah, today's zoological repertoire. In the Gilgamesh epic, which is older, there is a council of Gods – it is polytheistic. The Gods decide man must go – partly because he has been making too much noise! But there is one Goddess who betrays their confidence and disobeys the order to keep the plan secret. Against the will of the other Gods she secretly warns Utnapishtim. The Old Testament exegete commented that whatever one can say against Yahwe, the Father God of thunder and storms, the Jews have succeeded in incorporating in him the element of compassion. This is why their religion has survived while others have died out. It embraces the female elements. The words for 'Holy Spirit' and 'breath' are cognate with 'uterus'. Other female elements include the metaphors of the Prophets, for instance where God is described as a bride. I would welcome it if feminism and ecologism were united on this level.

AG You have written of two lines of thought in Christianity which could serve as environmental models: the Benedictine (or Cistercian), and the Franciscan. Do you see their ways of life as only viable for peripheral social groups, or perhaps as relevant to society as a whole?

CA I see them as a part of our culture, though not necessarily as something to be taught in school. Not as models for everyone, but as possible ways of life. The Benedictine profile is, let's say, more scientific than the Franciscan. It represents the idea of a *stabilitas loci.* Here, as in the seven-generation practice of the Indians, responsibility for a piece of land is given to a group of people for a lengthy period of time. They have to make do with the natural resources, and manage it. Their actions are guided by an order of life which, despite all the limitations imposed by its conditions, implies freedom. The freedom of the holy rule, as I'd call it, is not specifically Christian, but ecumenical. The individual in complete command of himself possesses, in Hegel's words, 'understanding of necessity', and recognises the possibility of developing a spiritual life in this freedom from desire, from ever increasing expectations. Then there is the Franciscan approach. You have to distinguish between the saint himself and the order – he was really an anarchist! This is really the only way of life which reproduces Jesus' refusal to care for the day in material terms. It has been practised by others in the East, Saint Sergius for instance. Such radicalism is alien to the Benedictines. In the long term, of course, it is scrounging – the Franciscans scour the Ark in search of food. Maybe this approach will be more practicable in the future. Probably both patterns of life will be needed. Here I see two valuable traditions.

AG I would like to move on to the question of regional identity. Love of one's home land, or one's native region, has usually been associated with conservative politics, and became synonymous with an aggressive, expansionist racism in the Third Reich. However, the concept of 'Heimat' has experienced a revival since the seventies, and its possible contemporary relevance has been explored in the context of a politics of individual emancipation, grass-roots participation and devolution of central powers. You have often been described as a Bavarian patriot. Do you use the term 'Heimat', and if so, what do you mean by it? Has it perhaps got a positive ecological component?

CA Yes, indeed. You may know my speeches – I have been invited to speak by all sorts of people, Bavarian local history associations close to the Christian Socialist Union, and so on. But *Heimat* without ecological consciousness is as difficult for me to envisage as ecological consciousness without a conception of *Heimat*. *Heimat* isn't necessarily localised, it's linked with the oldest senses, which makes it interesting. I once suddenly had the feeling of being at home in the middle of Yugoslavia, in a tiny, typically Mediterranean village, with nothing green about it. I discovered it came from the animal dung. In my childhood I lived in a small town where the farmers came in to market once a week with their cattle and horses, and the whole town smelt of their dung. The most primitive of senses, smell, is a good key, that's why local history museums are failures unless they can convey 'the cold sweat in the suits of armour', as it were. They may have the odd pitchfork, but the manure is missing, and everything is displayed in glass cases. The sort of local identity which goes with that is dreadful. Unfortunately you find it quite often in Bavaria. I see a minor secondary front for my activities in reminding the Bavarians what they're doing to themselves, especially the Upper Bavarians. It's atrocious. But perhaps it's not so different in other countries!

AG Of course! In Britain we have a flourishing heritage industry...

CA I'd be surprised if it weren't the case. But I did a lot of walking and hiking in and around the small Bavarian towns when I was growing up. My father was a professor at the Catholic Colleges in Freising and Passau, and we had many contacts in the countryside. He wrote enthusiastic articles, and a nice little book in the manner of Hofmiller, the well-known Bavarian writer and journalist (see Volke 1986) – in some ways better than Hofmiller – in the Bavarian tradition of writing on the local area. My father was very interested in local stories, he wrote a sort of *histoire de mentalités*. One book was *Wandern und Sehen* (Hiking and Seeing: Mayer-Pfannholz 1930). He travelled down the river Lech with the last of the raftsmen. I've got such things in my blood.

AG *Heimat* is defined by Ernst Bloch as what would be important for the future, that is not so much the natural environment of flora and fauna as the communities of people with whom one can identify, and who are perhaps less exploitative in their relationship with nature.

CA Yes, there is that, and secondly *Heimat* has implications for the whole question of energy consumption. We must stop flying all over the place, and so on. And then it's a question of fractal theory. How long is the coast of England? It's a matter of perspective! Take Gregory Bateson, whom I've read with great pleasure and gain. He says learning is difference, it's becoming aware of differences. Someone who has been in six different Hiltons, even if one of them is in Bangkok, has seen less of the world than the simple travelling journeyman in the old days whose knowledge of the world was limited to what lay between Böblingen and the next best small town. He travelled more slowly, looked more closely, learned to differentiate. The coast of England can be terribly long.

AG If you are walking a coastal path, for instance...

CA One can learn as much between Devon and Cornwall as anywhere else in the world. As a Catholic, for instance, I am rooted in my Bavarian homeland, and don't see why I should go shopping around in other cultures and religions, buying Hopi masks and doing Tibetan meditation exercises, which *I* wouldn't do anyway, because they're too strenuous! New Age religiosity lacks commitment and effort, that's another reason why I regard it with scepticism.

AG As a final point, I'd like to return to your most recent book, *Hitler als Vorläufer*. A characteristic of debate on the environment in Germany has been the frequent references to the Third Reich. On the one hand, writers like Max Frisch, Günter Grass, Peter Härtling and Wolfgang Hildesheimer have seen the ruthlessness of our scientific and industrial exploitation of nature today as standing in a continuum of destruction which found its most extreme expression in the Shoah. On the other hand, opponents of the environmental movement have sought to discredit it by pointing to ideologically dubious aspects of the critique of civilisation, and reminding us of the green elements in Nazi ideology. But your approach is rather different. Perhaps you could explain what link you see between the ecological situation and the Third Reich?

CA Very briefly: First, Hitler is naturally concerned with the question of sustainability. He defines his own aim as the 'preservation of the species'. He constructs a concept of nature which is one of his most reprehensible ideas. Nature is cruel – he calls it the 'cruel queen of all wisdom'. It pursues an aristocratic principle: the weak must go, only the strong survive. This vulgar Darwinism was current at the end of the nineteenth and the beginning of the twentieth century, both in civilisation criticism and in society itself. In the naval imperialism of the time, and in the accusation, which was widespread in Germany, that humanity has allowed itself to become a domestic animal. The 'cultural steppes' – I can remember terms like that from my childhood. Hitler's philosophy built on this way of thinking. But his really mean trick was to offer the Germans world domination in the guise of sustainability. Someone, he argued, must take this on, someone must return the world to its original, healthy state. (He devised a six hundred year programme to this end.) Somebody must see to it that the natural resources aren't exhausted, or rather that we are in control of the resources, and this is only possible if the most intelligent and efficient race takes control. Other peoples will have to limit themselves. That's how he formulated it, and that is the ambivalence, the monstrous ambivalence, of his success. An extraordinary success, if you consider that he was able to persuade people to go along with such a paranoid system.

Hitler saw the Jews as the great enemies of nature. That was the real basis of his murderous antisemitism. The Jew was a virus contaminating the world with the crazy ideas that one can protect the weak and that decisions can be reached by discussion – all absolute rubbish! In reality the Jew was much too clever not to recognise it. He was deliberately destroying the world, in order to gain control more easily. Instead of engaging in a fair fight, he attacked from behind, as it were. That is quite clear if you read, or plough your way through *Mein Kampf* (which is frightful). It comes across quite clearly. And this hasn't been done before, that's where I think I have done something new. The central importance of Hitler's concept of nature has been neglected. It is mentioned as a curiosity in all the great Hitler biographies, which are very commendable: Jäckel, Fest, Haffner and so on. But it wasn't possible to treat the subject in this way after the war, the time wasn't right. And it has been blocked unconsciously by contemporary historians. Unconsciously, because it's an extremely awkward subject for the future.

We have a culture which seeks on the one hand to preserve its achievements, while on the other it has a clear feeling its achievements are killing it. In order to resolve this dilemma, Hitler cut the Gordian knot. He said: we have a right to everything, to technology and its destructive potential, we must subjugate the peoples in the East and take possession of their territory. Otherwise civilisation will destroy itself. With 'Jewish impertinence', as he put it, they claimed they could do things better than nature.

AG And you see the possibility that somebody may come in future with a similar programme?

CA Whether one person or several I can't say. I am wary of making predictions! Hitler has remained very popular – in the Third World he is read avidly, from the understandable point of view: why should the pale-faced Nordic peoples necessarily be the master race? For example the Vietnamese Marshall Ky and Nkrumah in Ghana took Hitler as a model, and Hitler is read widely and remains popular in the Arab world.

I remember well the Mexican seasonal workers when I was a POW in the US. As we drove by they shouted 'Heil Hitler'! For them we were allies in the struggle against the Yankees. But I don't think the real danger for the future lies in petty chiefs of would-be great peoples. (There are a few of them in Yugoslavia.) It lies rather in a neat separation of Hitler's idiocies from his essential formula. If one substitutes 'social achievers' for 'race', and replaces Hitler's machinery of destruction with a subtler selection procedure, for instance using genome mapping, that's how I see the Gordian knot being severed in future. At any rate I reckon this is more likely than the Marxist vision of the future, which presupposes international heroism and acceptance that we *all* deserve a good life.

The interview took place in Munich, 17 September 1998, and was translated by Axel Goodbody

PART II

Environmentalism in Germany:
Political, Social and Cultural
Dimensions

CHAPTER 3

Anxieties, Visions and Realities: Environmentalism in Germany

Axel Goodbody

Germany was not initially at the forefront of the international environmental movement which began in the United States in the 1960s, spread to Europe in the early 1970s, and reached its high point at the Rio Earth Summit in 1992. The destruction of German industry during the Second World War had been followed by a lengthy period of reconstruction, known as the 'economic miracle', dominated by an ethos of scientific progress, technological advance and economic growth. Economic success became a substitute for discredited political aspirations. The longstanding German tradition of conservationism, largely but not exclusively underpinned by a powerful current of anti-modernism, had been seemingly definitively discredited by its assimilation into the Nazi ideology of Blood and Soil. The rise of the German environmental movement coincided with the first significant postwar recession, which sowed the seeds of disillusionment with economic growth and modernisation. As in the United States of America, where a popular environmental movement on the Left emerged from the Civil Rights movement and protests against the Vietnam War, environmentalism in Germany absorbed much of the protest potential of the declining Student Movement, which had transformed German society with its demands for political, social and educational reform in the late 1960s. The protest against nuclear power, which first came to a head in Wyhl (Baden-Württemberg) in 1974, differed from earlier German conservationist movements characterised by backward-looking concern for landscapes and communities threatened by industrialisation and secularisation in two ways: it was accompanied by demands for individual emancipation and participative democracy, and its critique of contemporary society was for the first time based predominantly on scientific evidence rather than subjective feelings.

Since the mid-1970s, environmental concern has played a more important role in public and private life in Germany than in most other European countries. The care which Germans take to sort and recycle household rubbish, the prominence of environmental arguments in German advertising, and the importance attached to 'nature' (onto which the values of harmony, beauty, goodness and justice are projected) as a social norm are distinctive characteristics of German culture. For many years Germany (alongside Denmark and the Netherlands) has taken the lead in initiating EU environmental legislation, it had (up to 1994) in Klaus Töpfer a Minister for the Environment with a uniquely high international profile, and it chose to mark the new Millennium with a major 'environmental' project, the Expo 2000, whose motto 'Man – Nature – Technology' echoed the subtitle of an influential book from the early days of the movement (Commoner 1971). Concern for the environment and preparedness to make sacrifices, in terms of individual freedom and taxation, seem to the outside observer typically German, despite the paradoxical absence of speed limits on German motorways, high levels of consumption, and the shrinking differential between public attitudes towards the environment in Germany and Britain since the 1980s.

The rise of German environmentalism, and the success of the Green Party in particular, have been explained as the result of social and political circumstances: relative affluence and the growth of the educated middle class in the 1970s favoured the new politics concerned with participation and the quality of life, while the government's ambitious nuclear power programme provided both an effective focus for protest and a paradigm for deep-seated anxieties about the social and political implications of technological development. The political vacuum on the left of the Social Democratic Party (SPD) which the Greens filled, and the configuration of parties whereby the Greens held a pivotal position between the SPD and the Christian Democrats (CDU) gave them disproportional influence. The environmental movement was also able to build on the political structures and expectations generated by the antiauthoritarian Student Movement. It gained new impulses and perspectives from the variety of feminisms which emerged in the 1970s, and benefited especially from association with the Peace Movement in the early 1980s, when Germany's position as the most likely theatre of operations in a possible nuclear war between the USA and the USSR generated public protest on an unprecedented scale.

However, the aspects of German *culture* which contributed to the intensity of the underlying environmental concern have often been overlooked. To begin with, the legacy of the past is visible in the desire to construct a new 'clean' German national identity, as evident in the adoption of progressive environmental policies in the late 1960s and early 1970s in both the GDR and the Federal Republic. Commitment to the environmental cause was, albeit often unconsciously, a belated substitute for resistance in the Third Reich. 'If in later years I devoted myself with an energy and a persistance which astonished many to opposing a different mass murder, the impending atomic catastrophe, it was a kind of substitute action. The impulse came and continues to come from a sense of ineradicable guilt at having failed back in 1942', the veteran Austrian journalist and campaigner Robert Jungk wrote in his autobiography (Jungk 1994: 176). Jungk had learned of the systematic extermination of the Jews in the Spring of 1942, but was unable to convince the newspaper editors in Switzerland, where he was living, of the reliability of his sources. It is striking how often German activists have referred back to the Holocaust as the origin of their commitment. The novelists Günter Grass, Wolfgang Hildesheimer and Peter Härtling have explicitly linked their opposition to the atomic bomb and nuclear power, genetic engineering and pollution, consumerism and materialism with the antifascist struggle, interpreting these negative consequences of modernity as manifestations of a destructive continuum rooted in Adorno's 'instrumental reason'. The earnestness with which German intellectuals have combated such aspects of modern society is understandable as a continuation (some might say a displacement) of the moral obligation to keep alive the memory of the destruction and suffering inflicted by the Nazis. In the works of the children's book writer Gudrun Pausewang, for instance, there are clear links between national guilt, personal motivation for writing, and the need felt to protect the environment. As the events of the Third Reich fade from living memory, the more extreme positions of the environmental movement have begun to look like neurotic obsessions. Michael Schneider already argued in the mid-1980s that the apocalyptic environmental pessimism of representatives of the West German Left reflected the traumatic disillusionment with socialism of a generation for which left-wing political ideology had served as a means of escape from the repressive structures of the Adenauer era (Schneider 1984).

Environmentalism also struck a deeper chord in Germany than elsewhere because it conformed to established patterns of national identification with the 'natural', contrasted with the 'artificial' civilisation of the traditional arch-enemy France, or the 'commercial'

empire of Britain. Nature is invested with particular symbolic significance in German cultural tradition: ironically, the Roman historian
Tacitus was the first recorded writer to envision the Germans as
simple, virtuous, uncorrupted tribes capable of defeating the Roman
legions (see Schama 1995: 75–87). Popular self-representations have
ranged from Grimmelshausen's unspoiled Simplicissimus to Eichen-
dorff's good-hearted *Taugenichts* (good-for-nothing). The
iconography of German nationalism, confirming an identification
with forests in general and the oak in particular – traditionally associated with liberty, vitality and permanence – suggests it is no
accident that forest dieback should have had such a powerful resonance for the Germans in the 1980s. German Romanticism and the
second phase of anti-Enlightenment thought at the turn of the twentieth century, the lifestyle reform movements, anthroposophy and
the Youth Movement, have also left a powerful legacy of holism and
pantheism whose influence is discernible in the environmental
ethics and aesthetics of the 1980s and 1990s.

This volume seeks to throw light on environmental concerns and
arguments in contemporary Germany, by bringing together findings
on German social values and political culture with studies of the
representation of human interaction with the natural environment
in fiction, essays, poetry, film and children's literature. The chapters
in Part III, on 'The Environmental Movement: Past, Present and
Future' are concerned with the roots of environmentalism in the
tradition of 'civilisation criticism', the reasons for the seemingly
dramatic paradigm shift to an environmentalist world view in the
1970s, the shift in the relationship between the Green Party and the
environmental movement from spearhead to an agent of weakening and splitting, the level of environmental concern (attitudes and
patterns of behaviour) in West and East Germany, and the 'ideological crisis' threatening the very future of environmentalism.

One focus of debate is the apparent decline of the environmental movement since the 1980s, and the prospects for the future.
Broadly speaking, Jürgen Hoffmann and Ingolfur Blühdorn argue
that ecological values are being abandoned in the West, and in
Germany in particular. This is a consequence of the dissolving of
the 'paradoxical synthesis' of the dynamic of individual, hedonistic
demands with the ecological ideology of renunciation, which characterised the seventies (Hoffmann), the 'unravelling of the
eco-ideological package' (Blühdorn). The environment is losing its
status as the potential source of a new social consensus, and ceasing
to represent a genuine alternative to the core values of capitalist
consumer society. While Blühdorn notes that the ideological certainties of the 1970s and 1980s have increasingly revealed themselves
as simplifications, and the Green Party has shown itself incapable

of tuning in to the *Zeitgeist* of youth in the 1990s, concluding there is no longer any 'space for ecological politics', 'no future for ecologist parties and organisations', Thomas Rohkrämer, Anja Baukloh and Jochen Roose sound notes of caution. They point to a continuity of environmental concern independent of political trends, which is borne out by certain statistics. This is likely to continue into the twenty-first century on a higher level than in most other Western European countries, even if environmental politics faces a pressing need to change in substance, style and ethos. The clash between environmental renunciation and the hedonistic lifestyle of the 1990s is one of several sub-themes also discussed in the 'cultural' chapters, which reflect on the role of books and films in framing and mediating perceptions of nature, naturalness and environmental crisis.

A second debate concerns the extent to which environmental concern and Green politics reflect realities or perceptions, scientifically demonstrable crises or human anxieties triggered by the violation of social norms and values. Hoffmann rehearses the argument that the origins, forms of expression and political significance of the international environmental movement derived from a shift of cultural and political values associated with the transition from industrial to post-industrial society. In addition to this, Rohkrämer and Blühdorn suggest, much as I have myself argued above, that the strength of the environmental movement in Germany and the particular concerns of the German public can be, at least in part, attributed to the way in which it tapped into German cultural traditions and responded to specifically German cultural concerns and identity needs. None of the contributors to this book would seek to deny the objective reality of an environmental situation urgently requiring action on a series of issues, indeed entailing a comprehensive rethink of our treatment of the natural environment. However, in studying the culture of environmentalism, the open-minded two-pronged approach combining 'nature-endorsing' with 'nature-sceptical' perspectives advocated by Kate Soper (Soper 1995) seems likely to yield important insights.

The chapters in Part Three of the book, which approach German environmentalism from the perspective of the disciplines of history, sociology and political science, speak for themselves. Before moving on to Part Four, which is concerned with literary and filmic discourse, it may, however, be useful to reflect further on the relationship between environmentalism and culture. Culture is defined anthropologically as the patterning of experience so as to give meanings to it and (normally) legitimate the goals of a given society. *Zivilisation* (social, political, economic and technical achievement) as well as *Kultur* (artistic endeavours and intellectual curiosity) are both commonly seen to perform a dual function, in *reflecting* and *directing* social values, including attitudes towards the environment. For cultural artefacts and events participate in the evolution of our understanding of nature through processes of conceptualising, stereotyping and imaging. The media, literature, film and art have all contributed in this way to the shaping of our perception of our relationship with nature today. Indeed, language itself, on which they all rely to a greater or lesser extent, is instrumental in 'constructing' the environment, as Rom Harré, Jens Brockmeier and Peter Mühlhäusler make clear in a recent study of the language of environmentalism, *Greenspeak* (Harré, Brockmeier and Mühlhäusler 1999). Research on the relationship between language and world view indicates that without actually determining our perception of reality, language invests objects with unreflected value judgements and gives our understanding of them a particular perspective. Words commonly have two functions, one through which they refer to reality descriptively, and a second which situates these realities in a system of social orientation and evaluates them. The term 'environment' is, in this sense, 'itself a blurred linguistic construction, a hybrid between nature and culture, matter and humankind, causality and morality' (185f.). The German word 'Umwelt' places man at the centre, and renders nature peripheral, more transparently than its English equivalent. For this reason, the environmental philosopher Klaus Meyer-Abich has called for its replacement by 'Mitwelt', implying a more physiocentric world view (Meyer-Abich 1990).

Metaphor plays a key role in the perception and comprehension of our environment, bringing otherwise unnoticed aspects of material reality to the fore, and delineating afresh, however vaguely, the boundaries of the phenomenon. The cultural geographer, William Mills, has suggested each age has possessed a dominant metaphor for nature (Mills 1982). Mills distinguishes between three periods in the history of the Western world characterised by the way in which they imaged nature. In the Middle Ages, nature was commonly seen as a book; in the Renaissance, it was believed to be organised in the same manner as a human being; in the Modern Age the most influential metaphor has been the machine. The book, the microcosm and the machine are not, of course, exclusive to their respective periods, but have served as central metaphors constituting the characteristic vision of a period. The choice of one metaphor rather than another is indicative of the needs and aspirations of that society, and the transition from one to another is associated with a reorientation of attitudes, indicating a fundamental change in the way we regard our position in relationship with nature. The machine image suggests human ability to control the world, in contrast with intelligibility and integration in the Middle

Ages and the Renaissance. It has encouraged us to see nature as a resource at our free disposal. But the gain in our control of the environment is achieved at a cost: nature is no longer resonant with meaning to us as in the Middle Ages, and no longer can we feel at one with it in the Renaissance manner. The logic of the machine metaphor is now perhaps unravelling, as its negative implications become apparent in the form of pollution and resource depletion, leaving behind feelings of guilt and anxiety. Other metaphors may come in time to be seen as comparatively more attractive.

Harré, Brockmeier and Mühlhäusler argue we need to find new central metaphors encompassing an ecologically sound relationship with the environment, and (following Meisner 1995) analyse the potential benefits and shortcomings of contemporary images for nature such as Gaia (the Earth Mother) from the point of view of their eco-friendliness. The personification of abstracta, allegories and terms with religious and cultural connotations may all have a part to play. Indeed, a multiplicity of different images may lend themselves to making us aware of their non-literal quality, and arming us against the shortcomings of any particular one. (Harré, Brockmeier and Mühlhäusler 1999: 177)

The term 'Waldsterben' illustrates the way in which the press can create and circulate new metaphors, encouraging readers to see things in a new light. *Der Spiegel* 'discovered' forest dieback in 1981, by publicising a phenomenon already familiar to experts, but suppressed by the authorities in order to avoid prejudicing the outcome of a dispute with Sweden over compensation for forestry damage originating from German industry (see Luhmann 1992: 9). The *Spiegel* series 'Das stille Sterben' ('The Silent Dying', echoing Rachel Carson's *Silent Spring*) made 'Waldsterben' a household word in December 1981. The choice of 'sterben', the word for animal and human dying, as opposed to 'eingehen', the term usually used for plants, suggested a bond between humans and the damaged environment, while the intimation that 'der deutsche Wald' (the German forest) was at risk evoked powerful cultural associations, and elicited immediate public response. The 'objective' official term 'neuartige Waldschäden' (new forms of forestry damage) was felt to be a euphemism. 'Waldsterben' emotionalised environmental debate and did much to fuel alarmist concern in the early 1980s – for a time, the annual *Waldschadensbericht* (official report on forestry damage) was allegedly taken more seriously than the budget – with sometimes dubious consequences. According to Marxist critics, the media in capitalist societies serve predominantly as agents of hegemony, affirming the status quo rather than initiating change, and reinforcing environmentally unfriendly modes of behaviour. By distorting or merely reifying reality, the press and television can of course effectively close off nature to us. However, the term 'Waldsterben' is evidence of the ability of writing and the potential of journalism to draw issues to popular attention and reconfigure the political agenda.

Policy makers too have opened up new perspectives with neologisms such as 'Umweltschutz', the compound for environmental protection first coined by the Brandt/Scheel Ministry of the Interior in 1969, and 'Lebensqualität' (quality of life), popularised by the Social Democrat Erhard Eppler (Eppler 1975). Environmental journalism, essays, and of course fiction and poetry, can be sites of imaginative fusion through which nature is reconceptualised, a new system of values negotiated, and a fresh ecological vision generated. Metaphors and images have the ability to engage the irrational and the unconscious through innovative links and associations. Reality can be re-mapped through narratives of hope and despair, and the skilful manipulation of temporal frameworks. Because of their focus on the medium, creative writers tend to be at the forefront of awareness of the shortcomings of received language. However, environmental writing in the broadest sense already validates the transition from the neutral discourse of science to that of morals, 'recruiting the force of scientific research findings to the task of persuading people to adopt a certain evaluation of some practice or programme' (Harré, Brockmeier and Mühlhäusler 1999: 50).

The chapters on literature and film in this volume seek to show the arts have been a part of public debate in Germany, reflecting and often seeking to promote the shift from a view of nature that is anthropocentric, resourcist, hierarchical, reifying and dualistic to one which is ecocentric, non-domineering, egalitarian, respectful, holistic, process-oriented and relational. Among those considered to represent creative writing were the veteran producer of TV nature programmes, environmental journalist, and – more recently – novelist Horst Stern, and the diarist, novelist and essayist Hans Christoph Buch, as well as the mainstream writers Günter Grass, Hans Magnus Enzensberger and Günter Kunert. Few German contemporaries may, however, be described with such justification as Carl Amery as 'environmental' writers. For Amery is unique in the energy, intensity and ability he has devoted to environment-related literary creativity, political activism and analytical reflection. As a person, he exemplifies underlying continuities with traditions of aesthetically and religiously motivated conservationist concern going back to the first half of the century. His radicalism, rooted in post-Second World War antifascism, is also representative of his generation. His writing mirrors many of the characteristics, including the tensions and contradictions, of the German environmental movement, in its blend of rational argument with ethical exhortation derived from a vision of the good (simple) life, of anthropocentrism with physiocentrism, of emancipatory political thrust with restrictions on individual self-expression in the name of the community and future generations. For all his left-wing sympathies, his environmentalism, like that of the Green Party, has arguably become a conservative force, seeking to slow uncontrolled change and limit the destructive potential of social and technological modernisation.

The chapter by Axel Goodbody on Amery is primarily concerned with illustrating the role played by the writer and intellectual in environmental debate in Germany. It strikes a balance between reviewing and assessing Amery's achievements as a novelist and as a Green thinker and essayist, and examines in particular his most recent book *Hitler als Vorläufer* (Hitler as Precursor: Amery 1998), from which an extract is translated for this volume. 'The Great Blind Spot', the final chapter of *Hitler als Vorläufer*, extrapolates a horror vision of a future in which the environmental challenge is met by a purely scientific and managerial approach, totalitarian domination and genocide from Hitler's social Darwinist *Weltanschauung*. Asking whether humanity can survive its own success as a species, and what the price in terms of human rights and human dignity will be, Amery pleads for a humanist, democratic, consensual, moral approach to the environmental crisis. The final section of the chapter on 'creative writing and the preservation of nature' sums up his contribution to the culture of environmentalism, and his achievement in informing, motivating and imaginatively empowering the public to face the environmental challenge, in the light of his own views on the social and environmental responsibility of the writer.

Jacquie Hope's chapter on poetry and prose fiction as examples of the cultural transformation of environmentalism in the GDR begins with a brief account of the environmental problems whose extent was only revealed when the government collapsed in 1989. She writes of the 'substitute function' of literature as an alternative medium of public debate in a society where statistics were censored and media debate suppressed. Ecological concerns were increasingly aired in the literary sphere from the 1970s on, when the international growth of environmental awareness coincided with tentative cultural liberalisation. Today, many of these texts appear derivative, didactic or artistically unfinished, but a handful possess lasting value as creative works of vision and sophistication.

The starting point of Rachel Palfreyman's chapter on German film is the curious absence of green films in Germany (feature films, as opposed to documentaries) comparable to Hollywood productions such as John Boorman's *Emerald Forest* (1984). The principal

German films concerned with human interaction with the environment are identified as those rooted in the ideologically burdened Heimat genre. In the 1970s and 1980s, critical Heimat films embraced environmental issues, but rarely foregrounded them. Environmental arguments compete with the Romantic legacy, apocalyptic visions and faith in the power of technology, reflecting many of the tensions and contradictions of German opinion today. Filmic conceptions of the environment, Palfreyman nonetheless concludes, have served to challenge viewers to reflect on fundamental alternatives to our way of life.

Finally, the role of culture in directing social values and the tensions between educational and aesthetic aims are addressed explicitly in Dagmar Lindenpütz's survey of children's literature as a medium of environmental education. In the 1970s, a new genre of children's literature concerned with the natural and social environment emerged in Germany, making readers aware of the dangers of nuclear technology, the extinction of species and global warming. The best of these books employ strategies of allusion and omission, using images and symbols to give readers space for their own interpretation. Some have become school classics. Lindenpütz identifies three broad interpretative models, which are matched in the GDR. Children's literature, she argues, has made a major contribution to ecological education in Germany, by helping young readers understand the problems of the world they live in, encouraging independent thought and stimulating the imagination.

In a volume of this kind, selection was inevitable. Space did not permit the separate treatment Austria and Switzerland deserve as countries in which the environmental movement has found distinctive political and cultural expression. Contributions on environmental education in Germany, on eco-theology and the role of the churches in fostering environmental debate, on the cultural foundations of the political parties (whose tax privileges in Germany permit them to play a role in public life unparalleled in Britain, ranging from research to the dissemination of ideas), on alternative culture, travel writing, theatre, art and sculpture were all considered, but could only have been included at the expense of others more central to the undertaking.

The anxieties, visions and realities by which the German environmental movement has been driven provide the common denominator of the studies presented here. The significance of individual aspects of nature in a given country is, as I have indicated above, culturally defined in a lengthy process whereby certain parts of the natural environment (for instance the forest in Germany, in Britain the countryside, or even beef) are invested with positive associations such as community, autonomy and permanence. In this process, creative writers and artists have played a role alongside politicians and journalists. This book may perhaps go some way towards helping British readers understand how Germans envision the environment. The differences in attitude between Britain and Germany have been highlighted over the last decade by the BSE crisis and the *Brent Spar* incident. Years of failure to acknowledge the dangers of BSE in Britain cannot, Richard Kerridge has argued, be understood in purely political or economic terms: the disease threatened to undermine our perception of farming as something other than industrial food production. EU measures to ban the import of British beef were consequently not experienced as a sensible precaution to prevent the spread of the disease, but as an affront to national sovereignty. Similarly, in 1995, while the Germans and Danes protested against the dumping of a disused oil rig in the Atlantic containing (as believed at the time) thousands of tonnes of toxic waste by boycotting Shell, the firm's action was defended in the British tabloids as our right in the face of foreign interference (Kerridge 1999). The German imagination was caught by the idea of the pollution of the high seas, the violation of the global commons. In Britain the countryside and animal rights have consistently ranked higher in public concern, spawning imaginative forms of protest against road building programmes and the export of live animals for slaughter. In a world where the 'reality' of the environment which we perceive is socially and culturally constructed, comparisons between languages, cultures and societies may perhaps serve a useful function in helping make us more aware of our own assumptions and their implications.

PART III

The Environmental Movement:
Past, Present and Future

CHAPTER 4

Contemporary Environmentalism and its Links with the German Past

Thomas Rohkrämer

In the Federal election in 1983, the Greens gained 5.6 percent of the votes. This was sensational: not only was it the first new party in the Federal Republic to jump the five percent hurdle and establish itself in the national parliament, the Greens also seemed to be 'a party completely new in essence' (Kelly 1982: 131). It saw itself as the 'alternative', a dynamic force against all the fossilised parties, which seemed to merely manage the status quo. The symbolic imagery of their first appearance in parliament strongly expressed this claim. First they boycotted the swearing-in ceremony of Chancellor Kohl to show their distrust. When they eventually entered, their appearance provoked an uproar. On this formal occasion, their sneakers, jeans and jumpers, long hair and beards were a powerful statement. They carried flowers and pine twigs, political badges and stickers. With these symbols they distanced themselves from traditional politics, associating instead with life, nature and the protest culture of citizens' initiatives and demonstrations.

Of course, there were critical voices who did not accept the claim to newness. Right-wing politicians saw the Greens as traditional leftists in disguise ('like a melon: green on the outside, red inside'), while left-wing observers frequently suspected the reemergence of a typically German romantic irrationalism. Many believed that the alleged 'flight from modernity' only promoted an ideology of stereotypes and resentment, instead of pragmatically managing the unavoidable necessities of modern life. Even some Greens started to search self-critically for 'eco-fascist' tendencies in their own movement. However, the majority stressed the novelty of the current situation and their thinking. Never before, they argued, had the threat of self-destruction loomed over humanity as a whole, never before had a science, that is ecology, shown objectively both human beings' negative impact on their environment and the way towards living in harmony with nature.

Many points about contemporary environmentalism are indeed so strikingly different from the past that one should be cautious about drawing historical parallels:

(1) The material conditions have fundamentally changed: the nuclear bomb has made human self-destruction an undeniable possibility, nuclear power stations or genetic engineering have opened wholly new dimensions to our dealing with nature, and the gradual emergence of a consumer and throwaway society in the 1950s was accompanied by a steep rise in the consumption of energy and other raw materials, in pollution and urban sprawl (Pfister 1996).

(2) Humankind has changed the world to such an extent that we live in a largely artificial world by now. The main challenge is not to preserve the relics of an allegedly untouched nature, but to manage a humanly shaped environment. Many of the dangers we currently face are no longer naturally given, but result from human activities. Technology may bring safety from natural disasters, but it simultaneously harbours its own risks (Beck 1986). This condition gives rise to unprecedented opportunities, challenges and dangers, ethical questions and responsibilities.

(3) The 'economic miracle' of the 1950s weakened reservations about modernity and revived the belief in progress. Economic slumps appeared to be a problem of the past, nuclear power promised to answer all energy needs, and a general rise in the standard of living appeased social tensions. According to opinion polls in the Federal Republic, whereas only 2 percent believed in 1951 that Germans were presently better off than ever before, the figure rose to 42 percent in 1959 and 62 percent in 1963 (Schildt 1995: 307). While an influential part of the German elites had traditionally been sceptical about modernisation, conservatism now merged with economic liberalism and came to accept the industrial market economy fully. Leading conservative academics like Hans Freyer, Arnold Gehlen and Helmut Schelsky did not question the 'technical age', the important conservative politician Franz Joseph Strauß claimed that 'to be conservative means to stand at the forefront of progress' (quoted in Wüst 1993: 40), and the radical conservative Armin Mohler (1974: 29) stated triumphantly: 'Leftists and conservatives have changed their roles. The left, which had prided itself

for so long on being the avant-garde, has now taken on the role of the luddite and thus of the latecomer in world history. But the conservative ... has been moved to side with industrial society, which he had distrusted for so long.' If German history had ever been distinguished by conservative distrust of American, British and French 'models' of capitalism and democracy, the Federal Republic had clearly 'arrived in the West' by the 1960s (Schildt 1999). The dominant attitude towards progress found exemplary expression in the words of the futurist Hasan Ozbekhan: 'Technologically we can do nearly everything. The question is: What should we do?' (quoted in Schulin 1979: 209).

(4) Environmentalism in the 1970s might have been particularly strong in West Germany, but it was clearly international in scope.[1] It seems to be a 'postmodern' phenomenon occurring in all highly developed countries. It does not take any particular anti-modernism, but rather prosperity for postmaterial values to gain a significant influence on political orientations.

(5) Despite some nostalgic notions, the mainstream of contemporary environmentalism has not harked back to some pre-modern past, but has tried to argue scientifically. Ranging from Carson's *Silent Spring* (1962) to the Club of Rome's study *The Limits to Growth* (1972) and *The Global 2000 Report to the President* (1980), the decisive environmental eye-openers were scientific studies, and it was at least initially widely believed that the science of ecology would show the way to a better future in harmony with nature.

(6) While the early environmental movement had been politically very diverse, the new left soon dominated. Nuclear power became the central issue, because it combined environmental concerns with a protest against large-scale capitalist enterprises and the fear of an overpowering state-capitalist system. The Greens as the main parliamentary representation of the environmental movement soon situated themselves left of the Social Democrats. Their programme was not limited to environmental issues, but also incorporated the call for radical disarmament, grass-roots democracy, feminism and social equality.

If all these points stressing the novelty of contemporary environmentalism are valid, why search for roots in the past? While the story sketched above is not wrong, it should not pass as the full story. How could environmental concerns emerge so quickly, if the ground had not been prepared before? Why did the new party emerge as a *Green* party, instead of giving socialist, feminist or peace issues centre stage? Why is environmentalism a bigger topic in Germany than in many other highly industrialised countries? The dramatic success of environmentalism in becoming a central political issue appealing to very different political quarters suggests that there must at least have been a strong cultural undercurrent making people open for these kind of concerns. The emerging Green movement of the 1970s could not draw on a conscious and living tradition; this was broken by the Third Reich and the cultural reorientation during the 'economic miracle'. However, the search for a less destructive modernity has a long history in Germany, as the ecological movement itself has discovered to its great surprise, and I shall argue that it has played an important role in the emergence of contemporary environmentalism. To make the point I shall first discuss past and then present environmentalism.

From the time when the discussion about Green links with the past started, historical views have changed significantly. Most importantly, a growing number of historians have come to question the widely held opinion that National Socialism was an irrational attempt to escape the modern world. The heated debate about the relationship between National Socialism and modernity/ modernisation has left many controversial points unresolved, but it has shown very clearly that most leading Nazis accepted modern technology or were even enthusiastic about it. In its self-presentation to the public, National Socialism also stressed its positive attitude towards technology. Instances are Hitler's use of aeroplanes for his election campaigns, the motorways, the programme to build a cheap car for the people or the propaganda about the wonder weapons which were to prevent defeat at the last moment. Nazi propaganda clearly used the latest technology, and the war demanded a full acceptance of industrialised society. Furthermore, racism was believed to be a scientific idea, and the alleged need for expansion was justified in economic, that is rational terms.

I do not want to suggest that National Socialism was a wholly modern movement. It was eclectic, drawing on many different traditions ranging from technocratic ideas to blood-and-soil mysticism. However, there is a growing acceptance that it existed within the framework of modern societies, that it showed many modern features and that its attempt to maintain power and achieve its central policy goals led to an all-out use of modern means. There was some agrarian romanticism, a revival of allegedly Germanic traditions and plans for an ecologically sound modernity,[2] but all this was never allowed to infringe on practical necessities. All in all, the technocratic emphasis on instrumental reason and technical fixes became increasingly dominant. While one can find some parallels between contemporary environmentalism and ideas in the Third Reich, these parallels are not more than is the case with many other convictions. Environmentalism did not protect Germans against the attraction of National Socialism, but neither did the belief in technology and progress (Rohkrämer 1999a).

While the relationship between National Socialism and contemporary environmentalism has been relativised, new connections between past and present have been discovered. Environmental history, which emerged in the 1970s as a reflection of contemporary interests, has revealed the environmental dimension of all history. Human beings have always had a strong influence on their environment. Traditional societies could cause great damage (e.g., through massive deforestation), but also succeeded in establishing sustainable usages of the countryside (e.g., through the three-field system). Industrialisation multiplied the human impact on the environment and could cause massive pollution, though until the 1950s still largely on a regional level. The single-minded emphasis on economic growth was by and large triumphant, but there were also massive conflicts and protests: town dwellers fought for clean drinking water and unpolluted air, farmers went to court to stop industrial emissions, which destroyed their harvest, and fishermen tried to save rivers from being turned into sewers. Governments and courts increasingly sided with industry, because the national wealth came to depend on it, but these protests never ceased, not even during the 'economic miracle'. The common image of the optimistic 1950s and 1960s overlooks many protests on a regional level, for example against water reservoirs, against the pollution of rivers and also against experimental nuclear power stations. In the election campaign in 1961, the candidate for the chancellorship, Willy Brandt, coined the slogan 'Blue sky above the Ruhr', calling for clean air for Germany's most polluted industrial area, and many experts became aware of the serious environmental damage caused by uncontrolled economic growth. All these initiatives neither achieved wider public resonance on a national level nor did they connect to form one united movement, but they do show an awareness of environmental problems, which the Green movement of the 1970s could draw on. There was interest in protecting nature and the environment, and environmental policies were actually initiated some years before the Green protest movement emerged. In his first government statement in 1969, the Chancellor of the new Social-Liberal government, Willy Brandt, spoke about the need for better protection, not only of animals and nature, but also against environmental pollution. At the same time, a department for environmental protection was founded on the initiative of some higher civil servants, and in June 1970 the cabinet passed a 'Crash Programme of the Federal Government for Environmental Protection'. While it should not be denied that it took citizens' initiatives and the protest culture of the 1970s to turn environmentalism into a central public issue, the ground was not unprepared (Brüggemeier 1998).

Conflicts over the environment are thus no new phenomenon, and neither are social movements with non-material goals. Social scientists speak of the 'new social movements' since the student rebellion, but the term 'new' is misleading. While it is probably true that the contemporary movements have had a bigger political impact, comparable phenomena have existed before. In particular towards the end of the nineteenth century, when it became obvious that Germany's future was that of an industrialised country, there was a whole wave of popular activities. Two organisations, the *Bund Deutscher Heimatschutz* (League for the Protection of the Homeland) and the Prussian state conservation agency, were founded, which called for the protection of unique parts of nature (Naturdenkmäler) and the preservation of historical monuments. They did not criticise industrialisation as such, but stressed the need to give non-material considerations (in particular aesthetics and naturalness) their due in all changes to countryside and cities. The life reform movement also emerged, which promoted a more natural life style within industrialised society, based on healthy food and natural remedies against illnesses, nudism and comfortable clothing, garden cities and a new educational style which aimed at harmony with the child's nature. The new Youth Movement, which achieved more freedom for the young generation to organise their own leisure time activities, is rightly known for its hikes in the countryside. The members of a Bohemian avant-garde also sought to find themselves by disregarding bourgeois norms. Some radical fundamentalists established communes, for example the highly successful 'Vegetarian Fruit Growing Colony Eden', to realise their desired lifestyle (Rohkrämer 1999b: 117–161).

These movements were no nine days' wonder. The Youth Movement multiplied in size during the Weimar Republic, agrarian communes existed under diverse political banners, and conservationism continued. While it shrank in size and influence during times of economic difficulty, it grew rapidly again between 1950 and 1965. In contrast to later citizens' initiatives, the different conservationist groups continued to work with the authorities, trying to influence their decision-making processes quietly with expert advice and warnings. However, they did move towards more ambitious goals. The initial attempt to preserve some small parts of an allegedly untouched part of nature or a threatened bit of national heritage was increasingly replaced by the demand for comprehensive land-use planning.[3] While acknowledging the inevitability of change as well as the need for industry and technology, they tried to give weight to aesthetic and biological considerations and prevent over-use (Dominick 1992: Part Three).

This more comprehensive environmentalism found its most noteworthy expression in the 'Green Charter of Mainau', which was formulated in 1961 by influential public figures from the political, industrial and cultural spheres. It expressed the warning that 'the basis of our existence has become endangered, because parts of nature essential for life are being polluted, poisoned and destroyed', and declared 'a healthy life in town or country appropriate for human beings' as a basic human right. Distancing itself from any romanticism, the Charter explicitly accepted industry and technology as 'indispensable prerequisites for our contemporary life', but demanded precise and enforceable planning to ensure enough space for recreation, a sustainable agriculture, the preservation or restoration of a healthy balance of nature, 'especially through the protection of the soil, the climate and the water', the preservation and sustainable use of all green areas and a stop to all avoidable damage to the landscape. As a practical measure, the 'German Council for Care of the Land' (Deutscher Rat für Landschaftspflege) under the patronage of the German President was founded, which comprised of ministers, higher civil servants, academics and a judge at the constitutional court. The panel did get involved in many environmental conflicts, most importantly over plans for an oil pipeline along Lake Constance, the canalisation of the river Mosel and the restoration of strip-mined land. The importance of environmental activities was officially recognised in the Council of Europe's decision to declare 1970 European Conservation Year.

In the realm of ideas, reinterpretations of critiques of civilisation in German history have also opened up new perspectives for hidden lines of continuity in environmental thought. Influential studies of the 1960s by Georg Lukács, Fritz Stern, George L. Mosse and Kurt Sontheimer had led to the widespread belief that critiques of civilisation were closely connected with the rise of National Socialism.[4] The argument was that a rapid and crisis-ridden industrialisation led to widespread antimodernist resentment, which in turn made the Nazis' antimodernist dream of a return to a truly German community of the people appealing. These studies focused on the political effects of critiques of civilisation, largely dismissing the content as an endless repetition of the ever-same stereotypes and negative feelings. In the last twenty years, this argument has lost much of its force. Firstly, the characterisation of National Socialism as antimodern has been, as I have shown earlier, seriously undermined, and secondly, the growing awareness of the dark side of modernity has opened historians' minds to the possibility that critics of civilisation might have had a point. While their political function should not be ignored, their warnings of negative aspects of progress and their search for alternatives are quite understandable and justified.

A careful look at critiques of civilisation reveals that at least since the end of the nineteenth century, when it became obvious that Germany was inevitably becoming an industrialised nation, the more influential thinkers cannot be categorised as dogmatic enemies of modern means or progress. They all saw negative aspects (that, after all, made them critics of civilisation), and many feared that modern technology might get out of human control – with potentially devastating effects. Yet most realised *nolens volens* that turning away from technology was not a realistic option. Even an extremist like the life philosopher Ludwig Klages, whose work has been ridiculed as 'Tarzan philosophy' by the undogmatic Marxist Ernst Bloch (1976: Vol 1, 65), accepted that technology, while being largely destructive at present, did have a positive potential, for example in educational films or in filters for purifying water. Instead of exploiting nature, a wise use of human means could, he believed, even 'fulfil the earth's wish for development' (Klages 1969: Vol 2, 1369) by enriching the environment with beautiful artefacts. His faithful disciple Werner Deubel went even further by urging 'technicians and engineers … to protect nature with technical means' (1934: 230).

Techno-enthusiasts defamed all critics of civilisation as dogmatic antimodernists, but in reality most accepted the difficult challenge of finding positive ways to integrate industry and technology into society, and they also developed intellectually stimulating concepts for achieving this aim. They knew that humanity could not turn back the clock and escape into the past, but aimed for another modernity. Their far from irrational vision was a reconciliation of their ideals with modern technology.

The social provenance of those who protested against an uncontrolled process of modernisation was also somewhat broader than has often been suggested. Of course it is true that, not unlike today, the early critics of civilisation were largely members of the educated middle class. They were able to make themselves heard, their lives were marked by a certain affluence, which allowed for the pursuit of non-material goals, and their incomes did not depend on jobs in industry. However, the appreciation of unspoilt surroundings and protest about environmental deprivation reached further and can even be found among the working class. The influence of Marxism, which regarded industrialisation as a necessary prerequisite for the improvement of material conditions and socialism, did not stop workers from answering in a questionnaire that they appreciated spending time in nature, missed a forest in their surroundings, or dreamt of an age with unpolluted air (Levenstein 1912: 354 and 360ff.). Equally, the socialist organisation 'Friends of Nature', which organised collective hiking trips and criticised the destruction of natural beauty through capitalist greed, shows that an appreciation of nature was by no means limited to a few well-educated romantics who wanted to selfishly defend a social position threatened by industrialisation.

In nineteenth and early twentieth-century Germany, a certain nostalgia was fashionable, and clichés about mass society, the horrors of the city and other alleged signs of cultural decline can be found quite easily. However, this should not disguise the fact that the more thorough critiques of civilisation did not argue stereotypically or hark back to an idealised past, but engaged with developments of their time, tried to formulate realistic alternatives and developed their views. Of course, no one critic of civilisation is exactly the same as another, but they may be classified within three ideal types, which consecutively dominated the public discussion. The limited space here only allows for the simplified sketch of an argument which I have developed in detail elsewhere (Rohkrämer 1999b). The early years of the German Empire were characterised by an attempt to turn technology into the obedient servant of bourgeois culture by making its use more ethical. These critics of civilisation appreciated the material achievements of industry and technology, but lamented the negative social and cultural effects like a decline in artistic production and communal spirit. In their attempt to overcome these 'side-effects', they placed their hopes in a renaissance of altruism and giving priority to high culture, which would put economic aims into their appropriate, that is subordinate place. Some saw modern means as a necessary evil only to be tolerated within the sphere of material reproduction, while others regarded them as essentially neutral means for all purposes. Industrialisation, they believed, had caused some negative developments, but only because humans were lacking the moral strength to use modern means in a benevolent way. The new riches had resulted in dangerous egotism and materialism, but used rightly they could also overcome social conflicts and be the basis of unprecedented cultural achievements.

Coming up to the turn of the century, a new generation of critics of civilisation came to distrust the former emphasis on anthropocentric ethics and cultural values. Not only did they believe that the appeal for moral strength had failed to make any difference, they even suspected dishonest double standards: for them the moralistic sermons had simply served as a cover, behind which business could continue as usual. As a consequence, some called for more moral rigour, but the truly radical forces went further.

Drawing on Nietzschean arguments, they came to distrust all ethics, trying to base their thinking on 'nature' instead. The different life reform movements and so-called life philosophers all attempted to find ways of establishing a more harmonious relationship between modernity and nature. The fear was that instrumental reason, technology and anonymous systems would suppress and eventually destroy both inner and outer nature. The human price for living in a mechanical age was criticised as much as the destruction of the landscape. Positively, their ideal was a world in which nature could prosper and realise its full potential in all its forms. This kind of critique could lead to a categorical rejection of modern means or to the utopian dream of fully reconciling nature and technology, but it mostly had a more moderate aim: to find a more natural way of life within the framework of modernity. Instead of increasing human domination over nature at all cost, modern means were to serve or at least respect life.

Already before 1914, this position attracted criticism. Can one really draw precise norms from nature? Is nature not ever-changing, taking on an infinite number of ever-new forms? Is not the will to increase our material well-being by technical means more natural than hiking through the countryside, contemplating sunsets or trying to conserve a static status quo? In the First World War doubts about 'nature' as a positive norm multiplied and spread. While for many the war of the trenches had revealed the destructive side of technology and the fragility of progress, it had also caused deep doubts about the power of ethics and an allegedly benevolent nature. In this bitter conflict, the belief that one could stick to tradition or 'nature' seemed to be naive, and after their defeat Germans had to give industry and technology unreserved priority, if they wanted to overcome their economic and political problems. In this situation, where modern means had revealed both their dangerous side and their inevitability with unquestionable clarity, the only realistic hope seemed to lie in a better organisation of industry and technology. Drawing on current economic ideas like Taylorism, the directed economy of wartime Germany, Fordism and technocracy, groups and individuals tried to design a perfectly organised industrial system which would not serve the individual interest of a group of capitalists, but Germany as a whole. They still voiced the same criticism as earlier critics of civilisation, but they felt forced to choose a fundamentally different remedy. Their diagnosis was no longer a lack of good will or naturalness, but a lack of competent planning. They blamed ignorant politicians and uncoordinated market forces. The rule of technical experts would overcome all problems. If the whole nation was organised like one efficient and well-balanced machine, it could, they came to believe, be productive without exploiting its human and natural resources or causing social and political divisions.

This sketch of different critiques of civilisation may well be too brief to be wholly persuasive. However, I hope it does at least serve to show that critics of civilisation did seriously engage with their time. They responded to material developments, learnt from criticism and failure, tried to improve their own concepts and drew on a wide variety of ideas to develop an up-to-date alternative to the status quo. However, of course, this does not mean that later critics of civilisation were superior or even had the right answers. Such a belief in a positive learning process did exist, but it was destroyed by National Socialism and the nuclear bomb. The technocratically oriented critiques of civilisation which dominated debate in the Weimar Republic connected easily with the extreme right and National Socialism, because they seemed to promise a strong and homogeneous Germany, ready to reemerge as a world power. However, while this vision of a directed economy, which would reconcile high productivity with respect of nature, social justice and a communal spirit, was an inspiration for National Socialism, the reality of the Third Reich revealed the hubris of such a belief in human planning and perfect systems. Furthermore, the unprecedented destruction of the Second World War and the threat of nuclear war made it all too obvious that humankind could not survive, if it did not find ethical ways of controlling itself and developing the will for a more humane world.

After 1945, critics of civilisation found themselves in a precarious situation. Ideologies typical for Germany came under critical scrutiny, ideological opponents could associate critiques of civilisation with National Socialism, the destruction demanded an unreserved emphasis on reconstruction, and the speedy recovery seemed to disprove all pessimism. On top of this, the visions of a better modernity had not only failed on a practical level, they had also revealed their theoretical inadequacy. Under these circumstances it comes as no surprise that many critics of civilisation tried to abandon their previous ideas.

Yet breaking with the past is not easy, one finds with the above mentioned Freyer, Gehlen and Schelsky. Their own belief, frequently accepted in secondary literature, that they taught conservatism to accept modern means by abandoning the tradition of cultural criticism (see for example Schelsky 1961: 9 and 32) needs serious qualifications. Firstly, the acceptance of industry and technology was, as we have seen, not really novel, and secondly, their acceptance was by no means unreserved. While it is true that they abandoned the hope for an alternative modernity, they did not come to share the belief in progress. To them the 'technical age' was not necessarily better than the past, but simply unalterable fate. To accept modern conditions was 'not a moral imperative' or a matter of preference, but 'a practical necessity'. 'Every action necessarily orientates itself towards progress ... to be effective at all. It has to do so, even if wants to oppose progress, or some of its tendencies, in order to curb over-hasty developments, correct wrong trends, or preserve threatened assets' (Freyer 1965: 88f.). While they thus accepted technological progress as the ground rule for all contemporary actions, they did try to preserve the autonomy of the 'whole human being' (Freyer 1965: 100, Schelsky 1961: 37, 40ff. [quotation]). At least for Hans Freyer the central and open question was not whether human beings could exist in the technical age, but whether they could exist '*as human beings*', who '*lead their lives*' (Freyer 1955: 229). While the dangers of turning outer nature into an artificial system did not come into view, there was at least a deep fear that humans would turn into robots.

However, there was also much harsher criticism of modern civilisation. In the first decade after the war, radical conservationists like Erich Hornsmann, Anton Metternich and Reinhard Demoll gave knowledgeable warnings of dangerous trends ranging from deforestation and steppe formation to the use of insecticides and population growth. In the novel *Der Tanz mit dem Teufel* (The Dance with the Devil: 1958) the founder of the 'World League for the Protection of Life', Günther Schwab, narrated a plot of the devil to destroy humankind by encouraging them to cause environmental disaster. The physician Bodo Manstein, who founded the 'Fighting League against Nuclear Damage', also wrote a scientific book about the dangers of radiation from nuclear testing, weapons and reactors (Dominick 1992: Ch. 5).

All these largely factual accounts contained more general remarks in which the influence of former critiques of civilisation was evident. However, there was even a small number of authors who could still be counted as critics of civilisation. Some texts of Karl Jaspers' or Ernst Jünger's, Martin Heidegger's *Die Frage nach der Technik* (The Question Concerning Technology, first printed 1954: Heidegger 1990) and Günther Anders' *Die Antiquiertheit des Menschen* (The Antiquatedness of Man: 1956) belong in this category, but arguably the most powerful example is Friedrich Georg Jünger's *Die Perfektion der Technik* (The Perfection of Technology: 1946). Its fundamental critique seeks to unmask all the promises of 'technological progress'. The text questions whether technology really saves time by pointing at the hectic speed of modern life, denies that it creates riches by pointing at the exploitation of non-renewable resources, and warns that all domination of nature also entails a suppression of our own human nature. The book culminates in the dark prophecy that nature will take revenge and destroy humankind, if our attitude does not fundamentally change.

The critique of civilisation is thus continued, but it has become more pessimistic. While authors of doom like Oswald Spengler and Ludwig Klages or fatalists like Max Weber had previously been exceptions, who were strongly criticised for their lack of a positive perspective, in the 1950s an apocalyptic tone prevailed among critics of civilisation. The above mentioned failure of all previous critiques, the smooth economic development and the feeling of isolation in a society with a growing consensus about industrial growth and a high level of consumption all led to their deep pessimism. The dangers were described in all clarity and sharpness, but the former visions of another modernity had disappeared, leaving behind only vague appeals for repentance and a complete change in attitude.

It was well into the 1970s before the new left started to pick up environmental issues, and even then most influential and substantial contributions came from other backgrounds. The director of the Frankfurt Zoo and television star Bernhard Grzimek had a strong impact with his calls for national parks, the ethologist and Nobel prize winner Konrad Lorenz gradually turned into a true critic of civilisation, and leading nuclear physicists started to question traditional science. While Joschka Fischer's book *Der Umbau der Industriegesellschaft* (The Conversion of Industrial Society: 1989) made him one of the few environmental experts among the leading Green politicians,[5] the influential best seller *Ein Planet wird geplündert* (The Pilfering of the Planet: 1975) by the right-wing politician Herbert Gruhl or *Ende oder Wende?* (End or Turning Point?: 1975) by the Social Democrat Erhard Eppler were published much earlier. Most leading environmentalist writers like Carl Friedrich von Weizsäcker, the Protestant theologian Günter Altner, the philosopher Klaus Meyer-Abich, whose father had already published on the philosophy of nature, or the writer Hoimar von Ditfurth had varying degrees of sympathy for the Green party, but they were not intellectually shaped by the new left. Their environmental concerns had other and older roots.

The writers who continued in the tradition of critiques of civilisation in the Federal Republic of Germany even established their own forum: the journal *SCHEIDEWEGE. Vierteljahresschrift für skeptisches Denken* (Crossroads. Quarterly for Sceptical Thought), founded in 1971, with Max Himmelheber and Friedrich Georg Jünger as editors. While Himmelheber financed the enterprise, the journal emerged intellectually from three discussion groups, one with Friedrich Georg Jünger as its intellectual centre, another, 'The Symposium – Discussion Group about the Responsibility of

Science', with the well-known theoretical physicist Walter Heitler, the biologist Heinrich Zoller and Joachim Illies as leading figures, and, most importantly, the 'Society for Anthropo-Ecology' around the historian Friedrich Wagner (Dietmar Lauermann and Jürgen Dahl 1989/90: 1ff.).[6]

Like most critiques of civilisation after 1945, the journal's initial programme (Friedrich Georg Jünger and Max Himmelheber 1971/72: 1–9) was marked by pessimism. It stressed feeling out of tune with the prevailing belief in progress, and sought to voice scepticism about commonly held assumptions and raise awareness about all the decisions society thoughtlessly takes. It emphasised the dangers of the 'risky venture of progress' (Abenteuer Fortschritt) and stressed the importance of feeling grief when breaking with the past, but did not suggest any alternative. In the first year, the journal contained many articles relevant for environmental thought, ranging from Friedrich Wagner's critical views on modern science to an article by the ecologist Gerhard Helmut Schwabe on nature conservation, and an attack on the Pope's glorification of the moon landing by the philosopher Franz Vonessen, but there was such a range of contributions that one cannot really speak of an environmental focus. Rather, the journal and environmentalism gradually started to realise their affinities. While early critics ridiculed the journal for its outdated tone (see Friedrich Georg Jünger and Max Himmelheber 1972/73: 5), a growing number of prominent figures from the environmental movement like Günter Altner, Carl Amery, Hans Jonas, Erhard Eppler, Klaus Meyer-Abich, Ernst Friedrich Schumacher or Ernst Ulrich von Weizsäcker gradually started to publish in it. At the same time, the editors came to give the journal a more precise and optimistic purpose and to define it in ecological terms. They not only opened their pages to critical articles on topics like nuclear power, they also published a 'Bussauer Manifest zur umweltpolitischen Situation' (Bussau Manifesto on the Environmental Situation). The document, which was co-written by the editor Max Himmelheber, warns that humankind heads for self-destruction and demands a fundamental shift towards a sustainable society based on 'ecological and human principles' (Jürgen Dahl et al. 1975: 469-486). By the 1980s, editorials explicitly stated the new direction as providing a forum for all thoughts which put 'ecological problems and the crisis caused by them' on 'a broad intellectual basis' (Max Himmelheber et al. 1984/85: 389, see also 'Zwölf Jahre Scheidewege' 1982/83: 713-715.

What is the historical significance of all this? The important influence of these traditions on the Green Party would reward study,[7] but is beyond the scope of this article. Politically, moderate and right-wing environmentalists played a significant role in citizens' initiatives and in the first regional environmental parties emerging in the late 1970s, but most of them left the Green party when it moved to include in its programme demands like social and gender equality, grass-roots democracy, gay rights and disarmament. Though a more conservative environmental party, the Ecological-Democratic Party (ÖDP) was founded in 1981, it has failed to become a political force.

However, environmentalism has a much broader base than the Green voters. Public opinion polls from the 1970s and 1980s show that environmental issues were generally given a high priority. At the end of the 1970s 20 percent stated that they wanted an environmental party, in 1984 55 percent could imagine that they would vote for an environmental party, and in 1981 51 percent rejected nuclear power (Noelle-Neumann et al. 1983: 105, 292–294 and 514), but the Greens never gained more than 13 percent of the votes in any major election, despite the fact that they were the only major party clearly opposed to nuclear power and strongly committed to environmentalism.

I do not want to belittle the importance of the Green party. On the contrary, it was able to gain an influence well beyond its limited electoral successes because its central political concern also appealed to non-Green voters. Environmentalism has become a topic no party can ignore, because there are many civil servants, non-government organisations, citizens' initiatives, activists, writers and, above all, voters, who are deeply concerned about this issue. Environmental problems are very real, but it also takes a culture open to them to recognise them and act. Critics of civilisation in Germany have clearly played a part in shaping a mentality open for environmental concerns. While their critiques have undoubtedly been marked by the prejudices and limitations of their time, and their social provenance, the broad scope of their concerns can be a useful reminder that environmentalism is not a purely scientific matter, but involves all aspects of human existence.

Notes

1. The difficulties experienced by *Bündnis 90/Die Grünen* in gaining votes in the new *Länder* is further evidence of the fact that environmentalism is not something peculiarly German.

2. Agriculture was occasionally promoted (although more in word than in deed), the new motorways were supposed to be in harmony with the countryside, and some of the plans for restructuring eastern Europe show an attempt to practise ecologically sound development. However, one must not overlook the fact that this was only done as long as it did not clash with German economic or military aims.

3. This move towards accepting necessary changes and aiming at a sensible and comprehensive management of the landscape was, at least in principle, already discernible before the First World War in the *Bund Heimatschutz* (Rohkrämer 1999b: 132ff.).

4. A variation on this topic is Herf 1984. For a critique of his view see Rohkrämer 1999a: 49f.

5. Another Green politician who gave environmental issues the highest priority was Rudolf Bahro, but he became increasingly isolated and left the party. Manon Maren-Grisebach's *Philosophie der Grünen* (Philosophy of the Greens, 1984) also failed to influence party discussions significantly.

6. In his main work *Die Wissenschaft und die gefährdete Welt* (Science and the Endangered World, 1964), Wagner not only formulated a critique of nuclear physics, but also attempted to show more generally how modern science had become so dangerous. This little known book offers not only a very detailed and knowledgable account of a large topic, but also puts forward a sophisticated and stimulating interpretation.

7. The heated debate whether Green thought can be classified as a 'conservatism of values', in contrast with the political 'conservatism of structures' (Erhard Eppler), reveals some awareness of this connection. However, the clichés which still dominate the general view of critiques of civilisation have largely prevented a more differentiated understanding.

CHAPTER 5

From Cooperation to Confrontation: The Greens and the Ecology Movement in Germany

Jürgen Hoffmann

Introduction

Ecological protest and State environmental policies became important political phenomena in Germany in the 1970s and 1980s. Whereas in the 1970s environmental protest was confined mainly to the extraparliamentary sphere, the founding of the Green Party in 1980 created a new institutional context. This chapter sets out to examine the development of the relationship between the Greens and the ecology movement in the Federal Republic of Germany. It examines the Greens' roots in the ecological movement and analyses their transformation in the parliamentary process and the day-to-day work of the Berlin Government. Taking the example of the red-green Federal Government since 1998, the individual successes and failures of green ecological politics are assessed. First of all, however, the origins, forms of expression and political significance of the green challenge are traced back to the shift of cultural and political values in postwar Western societies.

The Shift in Values and the New Social Movements

The emergence of the Greens is rightly seen as a result of the new social movements in Germany in the 1970s. These protest movements reflected a change in cultural and political values which, from the 1960s onwards, took place in the affluent societies of Western Europe. By values we mean the basic intellectual and moral convictions guiding people's attitudes and behaviour. A change in values involves a shift in the models for leading our lives, and in our conceptions of what is good, right and important. Changes in social systems of values can be due to the adoption of new positive values, to a shift in the ranking within the system of values, or to a decline of values. Research into change in values has been strongly influenced by the American sociologist Ronald Inglehart, who found in the early 1970s that materialist values had lost out in importance in Western postwar societies in comparison with post-materialist values (Inglehart 1977).

The structure of values of the war and pre-war generations was still very strongly determined by personal feelings of deficiencies and the resulting need for physical and economic security. Compared to that, the postwar generations grew up in a time of lasting peace and increasing economic prosperity. Under these different conditions, especially the young considered socio-political aims such as 'freedom of speech' and 'co-decision-making in politics and the economy' more important than a 'stable economy' and 'law and order' (Inglehart 1971). For Inglehart materialist values were the expression of a need for physical and economic security, whereas post-materialist orientations reflected needs to belong and be respected, and for intellectual and aesthetic satisfaction. He interpreted the shift in values as a generational phenomenon: since the new generations retained their newly acquired value orientations, as the older generations died out, and their place was taken by the young, a radical change of the system of values would inevitably result.

Inglehart's theses sparked off a controversial debate. Most researchers agreed that in the 1970s and 1980s the proportion of post-materialists was increasing in many Western industrial countries, whereas that of materialists was declining. At the same time it became clear that the post-materialists were far from constituting majorities in society. Even in countries like Germany and the Netherlands, where they accounted for a particularly high percentage, they did not form more than a quarter of the population at the end of the 1980s (Schmitt-Beck 1992: 529). Later research found that, in contrast to Inglehart's predictions, materialist and post-materialist values were actually intertwined in various ways. According to Klages, mixed types formed the most frequent type of values found empirically. Between 1970 and 1994, their frequency increased from 42 to 55 percent. Thus, the change in values was leading to a synthesis of materialist and post-materialist values (Klages 1993: 48). In individuals' value hierarchies materialist values like property, prosperity and career progression took high rank positions alongside environment-related or hedonistic values such as happiness and enjoyment of life.

In Germany, the new social movements marked the transition to a post-industrial society. Early in the 1980s, the term *new social movements* became accepted in academic usage. It covered the whole range of movements, from the anti-nuclear movement and the environmental movement to the squatter movement. These protest movements were very different in their organisational structures and forms of action, but they had one thing in common: their demands were related to the new politics. The adjective *new* distinguishes this type of movement in content and time from the older labour movement. While the labour movement showed a close link with class conflict, the new social movements focused on mainly ideational, non-material concerns. The new social movements are value communities advocating universal values such as peace and public assets like a clean environment, rather than individual interests (Veen 1988).

The new social movements consisted of the ecology movement, the peace movement, the feminist movement, the Third World movement, the alternative movement and the squatter movement. Their concerns ranged from the counter-cultural life styles of the alternative movement to the power-oriented political protest of the anti-nuclear and peace movements. Martin Jänicke has characterised all these movements as 'the self-help organisation of protection interests neglected in the power structures of industrial society' (Jänicke 1982: 72).

The new social movements began with the student movement of the late 1960s, which brought a sharp break with accustomed political topics and procedures. The mass demonstrations against the use of nuclear power in the second half of the 1970s were the climax of protest of the *citizens' action groups* and the *ecology movement*. The environment became a qualitatively new political and lifestyle paradigm in the 1970s. The costs of unlimited economic growth and of the destruction of the environment and the natural basis of human life had become increasingly apparent. Scientific studies of population development, pollution and the consumption of raw materials led to the prediction that in the absence of countermeasures the earth was heading for ecological disaster within the next hundred years (Meadows 1972).

The ecology movement (the term *Ökologiebewegung* gained general acceptance in Germany at the end of the 1970s) embraced the whole range of initiatives, action groups and associations working for the protection, preservation and improvement of the natural basis of life. It embraced three different aspects: traditional nature protection, pragmatic environmental protection and political ecology. They were united not only in questioning the traditional concept of progress, but also in a general scepticism towards technology and civilisation (Guggenberger 1975: 29). Since the late 1960s there had already been a State environmental policy in today's meaning of the word, but in the course of the 1970s the extraparliamentary ecology movement became the real motor of environmental politics. Increasing awareness of environmental issues led to the founding of a large number of local citizens' action groups. From the early 1970s on these fought against the, at times, obsessive aspirations of urban development, regional, transport and administrative planners. They had a very considerable impact, came up with memorable symbols and put the State authorities under growing pressure to justify their actions.

In the 1970s, the number of citizens politically engaged outside the political parties rose significantly, reflecting a fundamental change in aspirations for political participation.[1] The emergence of local initiatives was not only a reflection of increasing ecological awareness, but also a consequence of student protest in the late 1960s, which had expressed a strong sense of dissatisfaction with representative democracy and placed a new emphasis on the direct political participation of the individual. The *Bundesverband Bürgerinitiativen Umweltschutz* (Federal association of citizens' action groups for environmental protection, abbreviated as BBU) – a nationwide umbrella organisation of environmentally concerned citizens' action groups – was founded in 1972 as an association of fifteen regional groups. By the end of the 1970s it had an estimated 50,000 members, working in approximately 450 individual initiatives. In the 1980s, the ability of the ecology movement to mobilise the public declined. By the end of the decade the number of member groups of the BBU had declined to about 200. The BBU was an attempt to combine the principle of autonomous grass-roots initiatives with an effective central lobby. Although it could be regarded as a 'political ecology' organisation, the *Bund für Umwelt- und Naturschutz* (Association for the protection of the environment and nature, abbreviated as BUND), founded in 1975, belongs rather to the category of 'pragmatic environmental protection' (Rucht 1991: 344–8). In the mid-1980s, some 140,000 people and in 1992 already 200,000 people were organised in the BUND (Weichold 1993: 11, Rucht 1991: 347–358). Greenpeace, whose German branch was founded in 1981, also falls into this category, but Robin Wood, a grass-roots democratic organisation which broke away from Greenpeace in 1982, may be assigned to the political ecology camp. In the mid-1970s more than an estimated one million people were active in environmental organisations (Raschke 1993: 41).

Not only the environmental movement, but also the peace movement played an important role in politics. Primarily concerned with opposing the stationing of medium-range missiles in Europe, it peaked in 1983, when there were some 4,000 individual initiatives in which between three and five hundred thousand activists participated (Stöss 1984: 549). In 1983, the number of demonstrations in Germany increased by 74 percent over the previous year. However, in the following year it declined again by 19 percent.[2] At the beginning of the 1980s there was an ideological split within the German peace movement. One wing supported the concept of peace politics extending across the blocs, the other pursuing the unilateral aim of preventing the upgrading of the nuclear arsenal in the West. For the latter wing of the movement NATO was the driving force endangering peace, threatening a predominantly defensive Soviet Union. Within the Greens, Petra Kelly, Lukas Beckmann and Milan Horacek of the Czech civil rights movement belonged to the first wing. The second was mainly promoted by eco-socialist groupings within the Greens, who were quite influential up to the end of the 1980s (Knabe 1995: 1136f.).

By the middle of the 1980s, the new social movements had passed the peak of their ability to mobilise the public. Major protest events bringing thousands of people out on the streets were now the exception. Their concerns had been increasingly integrated into the party system – not least by the Green Party. In addition, the movements themselves revealed a growing tendency towards differentiation, institutionalisation and professionalisation (Rucht 1991). The individual aspects of the movements developed in different directions. When the rearmament debate was over, the peace movement no longer played a significant role as a mass movement in Germany. Its decline was accentuated by the East-West policy of détente, though it experienced a brief come-back during the Second Gulf War in 1991. The alternative movement took on the character of a movement concerned with practical issues of everyday life, especially in the subcultural centres of the big cities and in university towns. This became evident in subcultural and grass-roots cultural projects of all kinds – newspapers and cafés, bars and bookshops. The self-help and alternative sector expanded particularly between about 1975 and the early 1980s, remaining a decentralised phenomenon, despite strong tendencies towards professionalisation and institutionalisation. In the late 1970s, Huber (1980: 29) counted around 11,500 individual projects pursued by some 80,000 activists within the alternative movement. In the mid-1980s there were an estimated twelve to fourteen thousand projects with about a hundred thousand participants (Brand et al. 1986: 255). The squatter movement reached its peak in the second half of the 1980s. The centres of this movement were Berlin and Hamburg. In the 1980s, the autonomous feminist movement also lost the vigour it had possessed in the 1970s. This was not least due to the fact that questions of emancipation and equality had been taken up by the Greens, and subsequently by other political parties.Today they have long since become political and social reality. Thus the issue became the common property of all the parties. The Third World initiatives and aid action groups were less concerned with East-West deténte than the North-South relationship, focusing on economic and cultural questions as well as the political and military complex.

A Difficult Relationship: The Greens and the New Social Movements

The Founding Period 1977–1983: Multi-Coloured Protest in Party Guise. The Greens Between Movement and Parliamentary Party

The founding of the Greens was closely linked with the new social movements. The emergence of the Green Party, which began with the first Green Lists in 1977, was a direct result of the crisis of the anti-nuclear movement after the confrontation between militant opponents of nuclear power and the police on the construction site of the reactor in Grohnde (Lower Saxony) in the same year. Within the movement people began to think. An increasing number of anti-nuclear activists felt that occupying construction sites was an inappropriate form of action, because it led to provocations and violence on both sides. Moreover, the violent actions of a few discredited the demands of the citizens' action groups as a whole. Therefore, from 1977 onwards a part of the movement favoured a parliamentary representation of environmental concerns. The Green Lists sought to block large-scale technological projects such as nuclear power stations with the help of the parliamentary platform.

Besides the citizens' action group movement and the new social movements a number of further political and social groupings went to make up the Greens. In the first decade leading representatives of the so-called *K-groups* (the communist groups *Kommunistischer Bund* [KB], *Kommunistischer Bund Westdeutschlands* [KBW] and *Kommunistische Partei Deutschlands* [KPD]) played an important role. Subscribing to a policy of entrism, they saw in the developing ecological lists and parties a potential vehicle for socialist ideas, and considered them an intermediate stage on the way to a strong socialist party in Germany. The KB (Communist Association) in particular sought a new political orientation through a policy of broad alliances and commitment to the green-alternative movement (van Hüllen 1990). Individuals and organisations of the undogmatic New Left, among them the *Sozialistisches Büro* (Socialist Bureau, SB), soon showed an interest in the Greens too. The undogmatic left-wing political groupings in the Greens also included anarchist groupings like the *Spontis* and the *Stadtindianer* (Spontaneous Activists and Urban Indians). These distinguished themselves clearly from the centralist structures of the *K-groups*. Whereas the *K-groups* had a Marxist, economically determinist view of society, the anarchist groupings demanded 'subjectivity and the development of sensuality, feelings and individual autonomy' (Göbel and Guthke 1979: 869). Social Democrats who were attracted to the Greens in protest against the defence and nuclear energy policies of the Schmidt government were also influential.

Several very different political streams thus converged to found the Green Party. Apart from the green of the ecology movement, there were red (socialist) and purple (feminist) groups, as well as a multi-coloured protest which 'reflected the multiplicity of alternative approaches united only in expressing the social need for cultural pluralisation in their respective spheres of activity' (Raschke 1993: 41). Raschke describes the founding consensus of the Greens as leftist, ecological, libertarian and solidaristic.[3] Social justice, non-violence, decentralisation, equal rights, direct democracy, political participation and human rights were topics which also played a role in other political ideologies – for instance in socialism or liberalism – but this combination of them was new. With this mixture the Greens found support not only among voters seeking freedom and participation for the individual, but also among those who no longer believed the established parties were in a position to respond to the intensified political and socio-economic problems.

The founding period of the Greens lasted from 1977 to 1983. It started with the formation of green, multi-coloured and alternative lists on the *Länder* level and finished with admission of the Federal party *Die Grünen*, founded in January 1980, to the German *Bundestag* on 6 March 1983. The period was characterised by rapid growth of the Greens in the *Länder*. Their advancement proceeded parallel to the extraparliamentary mass protests against nuclear power stations and the stationing of medium-range missiles in Germany. The upswing of the Greens was primarily a result of 'the success of the movements of which the new party was considered the parliamentary representation' (Kleinert 1992: 293). The Greens took up the new social movements' range of themes, thus emphasising their strong connection with extraparliamentary protest. Whereas the old politics concentrated mainly on economic growth, price stability, defence capacity and a stable currency, the proponents of the new politics in the party system demanded economic activities be subordinated to ecological considerations. In addition, they called for unilateral disarmament, solidarity with the Third World and the protection of religious, ethnic and sexual minorities. Further demands were withdrawal from nuclear energy and equality between women and men.

The founding of the Green Party was thus not least a reflection of the realisation that social movements have only limited possibilities of influencing politics. Being excluded from the process of parliamentary debate and the formulation of legislation, they need parties to pass on their ideas. For this reason, the pragmatic wing of the movement supported a parliamentary representation of environmental issues. Participation in the *Bundestag* and *Bundesrat* offered greater scope for political influence and better access to information, and held out the promise of a solid organisational and financial basis. However, considerable resistance had to be overcome before the party was founded. Had not the environmental initiatives emerged precisely in protest against the parties? Was not there a danger that the Greens, as a new player in the parties' game, would gradually lose their critical stance towards the system and their radical opposition character? The party founders sought to prevent this through the construct of a 'movement party', according to which the Greens were to be, as it were, the parliamentary action committee of the various protest groups. They were to win people over to the movement's concerns and to provide the extra-parliamentary groups with information otherwise inaccessible to them. At the same time, the party protagonists were to stay active in the movements. In order to prevent their creeping adaptation to the established system in everyday parliamentary politics, elements of the 'soviet' (council) model of democracy were combined with a grass-roots democratic conception. Rejection of a politics driven by economic rationalism went hand in hand with a strong preference for direct democratic decision-making processes and unconventional forms of political participation. Behind this lay the concept of a 'movement party', which was, according to Raschke, 'in terms of the individuals involved and possibly also of its organisation, interwoven with the corresponding movements, and tied in with them in political interests, legitimation and mobilisation' (Raschke 1993: 499).

However, the structural differences between the Greens and the environmental movement were already obvious at this early stage. In contrast with the protest organised by the movements, the Greens were never a single-issue organisation. They rather combined emancipatory socio-critical demands with the principle of ecological responsibility. The Greens did not manage to separate the ecology protest from hedonistic motives: on the contrary, the dynamic of individual demands and the ideology of renunciation formed a paradoxical synthesis. Despite these opposing tendencies and the advancing parliamentarisation of the Greens, from 1978 to about 1983 their relationship with the ecology movement was characterised by close cooperation. In a nutshell, this period was characterised by the quest of the extraparliamentary opposition and citizens' action groups for a parliamentary platform for their political concerns, and a balance between extraparliamentary activism and parliamentary initiatives.

On the Way to an Established Party: The Greens Between Realos and Fundis 1983–1990

Despite occasional setbacks, the history of the Greens in the 1980s was an astonishing success story. In Federal elections up to 1987 the newly established party (founded only in January 1980) showed a trend of continuous advancement from 1.5 percent in 1980 to 5.6 percent in 1983 and to 8.3 percent in 1987. In 1987 the Greens increased their share of the second vote more than any other party. In the European elections of 1984 and 1989 the party also achieved very good results (8.2 percent and 8.4 percent). In 1983 the Greens were the first new party since 1953 to gain admission to the German *Bundestag*, thereby changing the party landscape. Admission to the German *Bundestag* marked the beginning of a second phase in the development of the Greens, which lasted until German reunification in 1990. During this period the party found itself in a situation favourable to its thematic concerns. After the reactor catastrophe in Chernobyl in 1986, the environment rapidly gained in importance. All party programmes were now given an environmental dimension. In 1986 eco-political responsibilities were united in a new Ministry of their own. The Greens, however, were unable to make use of this political tail wind; their energy was expended in tiresome internal struggles which almost split the party. The endeavours of the pragmatic wing to transform the Greens into a coalition partner for the SPD were to account for this. In 1985, the first red-green coalition was established in the *Land* Government of Hesse, but it broke up over the question of nuclear energy already in 1987. In 1989, there was another red-green coalition in Berlin which did not last very long either. Nonetheless, all the endeavours of the radical wing to stop the course of parliamentarisation failed.

After admission to the German *Bundestag* in 1983, the party gradually began to emancipate itself from the environmental movement. This was the beginning of a new period characterised by greater restraint in the relationship between the Greens and the ecology movement. Already in the autumn of 1982, after the *Land* elections in Hamburg and Hesse, the question of Green participation in government in the shape of either toleration of an SPD minority government or a formal coalition was on the agenda. The focus of green politics shifted with the accession of the Greens to *Land* governments. (In December 1985, Joschka Fischer was sworn in as the first green environment minister in Hesse.) At this time the political influence of the ecology movement began to decline. Its mobilisation capacity suffered as its ideals and demands were taken up by other political protagonists. During the protests in Brokdorf and Wackersdorf in the mid-1980s the anti-nuclear movement experienced a short high, but after that its activities, and those of the whole ecology movement, shifted to selective, decentralised interventions. The central topic of the ecology movement in the 1980s was the problem of forest dieback.

As a result of the growing success of the Greens, the themes of the movement were integrated more and more into the party system. Ecology politics thereby became a long-term political factor, but it left the extraparliamentary stage bit by bit and became a part of normal parliamentary activity. In the course of this trend, the Greens necessarily began to distance themselves from their movement roots. They became an independent political factor with tensions in their relationship with the extraparliamentary protest groups. Depending on the point of view, the practical political activities of the Greens can be labelled as either a betrayal of their original principles, or as successful, client-orientated politics within the framework of the politically possible.

The declining relevance of the movements arising from the parliamentarisation and professionalisation of the Greens had soon been recognised. 'Since under normal conditions in systems of parliamentary representation interests can only be carried through if they are selectively bundled ... pragmatic green politics in parliament entails both integration and exclusion. In this respect, the Greens have actually contributed to the weakening and splitting of a movement which is for its part unable to obligate the party to its principles by means of pressure from outside' (Stöss 1987: 295f.). However, the new social movements not only mobilised voters for the Greens. These rather competed with the SPD for the voter potential of the movement sector. As Pappi has shown (1989: 25), in the 1980s, the Greens relied heavily on the movement sector, but in comparison with the SPD they failed to profit from this sector optimally. Only 31 percent of the consistent supporters of the anti-nuclear movement voted for the Greens, whereas 50 percent of them voted for the SPD. The supporters of the feminist movement and the peace movements also preferred the SPD.

The relationship between the new social movements and the Greens also gave rise to quarrels over the direction within the party in the 1980s. The political realists had always had an ambivalent relationship with the movement. On the one hand they emphasised that the Greens needed to take their relationship with the extra-parliamentary movements as a touchstone in important decisions, because really significant changes could not be brought about by mere parliamentary majorities (Fischer 1984a: 133f.). But on the other hand they were the first to recognise the limits of reference to protest movements. A party, they stressed, was subject to a different political rhythm, to a different logic of alliances, to other forms of organisation and to different phases of activity. 'Protest movements need to attack the status quo from their respective point of view and to try to destabilise it, this is the source of their creativity, and what makes them indispensable in a democracy. Parliamentary parties, on the other hand, strive for hegemony, i.e., they seek to institutionalise their aims in a new social and political consensus' (Fischer 1984b: 32). Political parties could not serve as the political action committees of their respective movements.

This, however, was the precise intention of the radical ecological wing around Jutta Ditfurth, who raised close relations with the movements to the *raison d'être* of the Greens. MPs were to be the mouthpiece or spearhead of the movements. 'With one leg we wanted to stand firmly within the citizens' action group movement and ... play an active part there. With the other "parliamentary" leg we wanted to be the "yeast in the dough" of the established parliamentary parties, we wanted to introduce new ideas in parliament, and at the same time be a forum and mouthpiece for the political ideas and wishes of the citizens' initiatives' (Hasenclever 1992: 104). Anchoring the Greens in the movements was to guarantee the connection with the network of movements and assure that the parliamentary level of environmentalism did not become independent. The Party was to be, in a popular phrase, the parliamentary 'free' leg (*Spielbein*), while the extraparliamentary movements were the political 'pivot' leg (*Standbein*) (Vogt 1990: 172).

After 1983, the Greens could then still be regarded a movement party, but only 'in a general sense, meaning it was oriented towards the movements from which it originated, among other things'. The term 'movement party' had become practically meaningless. Raschke gives as the most important reason for the growing distance between the Greens and movement politics the logic of action determined by party rivalry: 'The tension between a movement party true to its principles but unelectable, and a party seeking to optimise votes, but not entirely without principles, is such that the compulsion to choose between them may be glossed over in words, but hardly in practice. If one is not ready to accept the special logic governing the actions of political parties, one would have done better not to withdraw resources from the movement and invest them in the open-ended experiment of a party' (Raschke 1998: 39f.). In addition, a party's orientations are inevitably related to the results of the elections, i.e., success and failure are calculated differently from a movement's. Another reason for the failure of the movement party is that a movement cannot be sustained indefinitely. As a result we can say that the close link between the party's and the movement's politics, which initially had been the goal of both sides, failed.

Farewell to Principles. The Greens' Transformation into a Governing Party

The Greens had made great strides towards establishing themselves in politics in their first decade, but German reunification changed the political situation fundamentally (Veen and Hoffmann 1992). This was the starting point for a third phase in the development of the Party. During this period the Western Greens were thrown out of the *Bundestag* after the first all-German elections on 2 December 1990. They merged with the Eastern Greens the day after the election and with the larger *Bündnis 90* (Alliance 90) in May 1993 to form the all-German Green Party. The first half of the 1990s was characterised by a reduction of conflicts within the party and by an increasing personalisation and professionalisation of green politics. At the same time, most of the grass-roots democratic instruments, like the principle of rotation, were hollowed out step by step until they were finally abolished. Thus faded the last remaining symbolic associations with the new social movements. The departure of the eco-socialists around Thomas Ebermann and Rainer Trampert from the Party in Spring 1990 was symptomatic of this major shift. In May 1991 about 300 supporters of the radical egologist Jutta Ditfurth then left the Greens in order to found the *Ökologische Linke/Alternative Liste* (Ecological Left/Alternative List). They sought to achieve 'a closer network and a better coordination of the work of the eco-radical, socialist, autonomous and feminist political groupings' (Ditfurth 1991: 316f.). However, the *Ökologische Linke* never managed to be anything more than a splinter group.

In May 1993, the East German *Bündnis 90* merged with the Greens to form *Bündnis 90/Die Grünen* (Hoffmann 1998). In the general elections in 1994 the party rose to be the third political power in Germany. In 1998, they joined the SPD in the Federal Government and underwent a significant change concerning their political role. The relationship between the Greens and the new social movements also had to be redefined (see Schmidt 1990: 1-3, Rucht 1987, Stöss 1991). However, the movement's expectations were based on a political misunderstanding. Many activists of the anti-nuclear movement hoped that with the support of the Greens they could attain their main objective of a rapid withdrawal from nuclear energy. During the 1990s, there had only been a few protest events mobilising significant numbers, the most important of them being the conflicts over the transportation of nuclear waste in high-security *Castor* containers. Here the militant autonomous scene moved increasingly into the limelight. Its role can be compared with that of the *K-groups* within the anti-nuclear movement in the 1970s, who played a key part in the escalation of violence at the reactor construction sites in Brokdorf and Grohnde. Just as for the militant *K-groups* of the 1970s, the activities of the anti-nuclear movement served as a vehicle for the Autonomous groups to express their anger against the system (see Blank 1998).

The inclusion of the Greens in Federal Government was anything but the anticipated signal for withdrawal from nuclear energy. Disillusionment soon spread among the movement. The citizens' action groups, which after almost twenty years in opposition now demanded the practical realisation of their political demands, met with cautious restraint on the part of the Greens, who were bound by governing party discipline. Many environmental groupings and associations no longer regarded the Greens as their ally, because they had relinquished the goals of the movement. Thus, the BUND repeatedly accused them of having given up their objective of withdrawing from atomic energy and of attaching greater importance to the stability of their coalition with the SPD than to their political identity. They maintained the Greens had given up their political identity for the sake of peace within the coalition, and that they were defending the status quo and had lost interest in translating their original political demands into practical action. The results of the talks between the Government and representatives of industry on withdrawal from nuclear energy in June 2000 have done nothing to change this. As the agreed operational life for each nuclear power station is 32 years, the so-called 'withdrawal' may rather be described as a gradual phasing out of nuclear energy. Since electricity generation can be handled flexibly, i.e., transferred from one station to the next, it remains open when the last nuclear power station will be taken off the grid. Thus, on this question the Greens have once again had to swallow a bitter pill.

Withdrawal from nuclear energy, the overriding concern of the new social movements in the mid 1970s, and later one of the principal demands of the Greens, is symptomatic for the failure of efforts to achieve the movement's objectives through either mass protest or party politics. Roth (1998) reaches a sobering conclusion when he notes that the Greens, once a part of the democratic motor of the new social movements, have today become a retarding factor and a brake on their activities. The Greens and the new social movements thus now face political ruin. Whereas the anti-nuclear movement is eager to establish new alliances with partners from the labour unions and the churches in order to pursue their aims, the Greens fear further losses among their hitherto loyal voters. They face a dilemma which is hardly soluble. If they keep to their traditional positions they can count on the – admittedly shrinking – support of the left-wing alternative milieu. However, more than 90 percent of the population will be against them. If, on the one hand, they moderate their political demands, they will continue to lose core voters in the movement and the alternative milieu.

It becomes obvious that the close dovetailing of the new social movements with the Green Party must soon come to an end. As the two have different structures and as they pursue divergent patterns of communication and interaction, their relations can only be troubled ones in the future. Over the last two years, the main disappointment of activists of the movement has been with the Greens' difficulties in gaining the acceptance of the SPD for their concerns. However, the reasons for the development depicted above do not lie exclusively on the side of the Greens. The political context has changed radically since the beginning of the 1980s. The Greens' success in the 1980s had only been possible because the established parties were slow to recognise the new topics and problems and the change in social needs, and failed to respond to them effectively to begin with. These new topics were brought into the political system by the Greens, and by the second half of the 1980s they had acquired a prominent place on the political agenda. At the beginning of the 1990s, there was, however, a reversal of trends back to materialism. Post-materialist values and 'soft' political topics receded into the background again. It thereby became clear that social values are closely tied to general political and social conditions. Whereas in the 1980s optimistic assessments of the economy had been accompanied by an increasing orientation towards personal development and the quality of life, by the early 1990s, when a sceptical to pessimistic evaluation of the economic situation prevailed, even the young generation turned back to traditional values and the 'old' politics (Veen and Graf 1997: 15).

The debate on globalisation, Germany as an economic location, mass unemployment and increasing social problems replaced ecology at the top of the political agenda. The Greens have not been able to find an answer to this change of paradigm so far. Their declaration of principles is the oldest of all the political parties in Germany, and their programme is out of sink with political trends. Only in the sphere of ecology do they possess a clear profile of special competence. However, even the Greens' voters have become choosey. In the face of the changes in the political agenda in the 1990s, the Greens are being forced to redefine their profile. Political analysts have advised them to drop 'loser topics' like ecology and take up positions on future-related topics if they want to be successful again.

The End of a Partnership: The Greens and the Ecology Movement Facing New Challenges

Increasing awareness of environmental issues and growing discomfort with representative democracy had been the driving forces of the ecologically motivated grass-roots protests. With their new and unconventional forms of protest, the Greens and their supporting movements contributed to an increasing readiness of citizens to participate in politics and to the acceptance of new political objectives in the programmes of other political parties. Especially in the fields of the environment and feminism, new topics found their way onto the political agenda. In some cases they have become the subject of legislation. However, faster than expected, the Greens have become part of the established political system. With their integration into the political party landscape, they – more or less unwittingly – served to reinforce the political institutions. They bundled the extraparliamentary protest, provided it with a political platform and integrated it into the political system. Thus, they helped to consolidate existing political structures and widen the basis of legitimacy of democracy in Germany. The Greens and the new social movements have thus strengthened the system they once fought. The successful embedding of the Greens in the organised rivalry between the political parties is at the same time evidence of the innovative and integrative capacity of the political system. 'The elastic party system in Germany has been able to defuse the once radical and utopian models of the Greens and to integrate them largely into democratic structures. Thus, the party system has been strengthened and the Green Party has gradually been modified – and in a way become part of the establishment' (Jesse 1992: 57).

The mobilising energy of the Greens and the new social movements has, however, dissipated as their ideas and demands were taken up by other protagonists within the system. Since the middle of the 1980s, all political parties have integrated environmental topics into their programmes, which has made the Greens lose their distinctness. The very success of the Greens and the ecology movement in providing the new political topics with a platform and getting them a hearing has become a major problem. It is no longer enough merely to point to political deficits and shortcomings. The symbolic politics of the early years has reached its limits.

The Greens now face serious problems regarding their future. Their recruitment basis is melting away; especially young voters, hitherto their loyal supporters, are deserting them in crowds. They are now regarded as an ageing party associated with a particular generation, suffering from communication problems with the younger age groups (Bürklin and Dalton 1994). Against this background the question has been raised whether the Greens might be nothing but a temporary project for the protest generations of the 1970s and the 1980s. According to their former mentor Udo Knapp, the Greens' left-wing opportunism has cost them the opportunity of becoming the 'popular party of ecologically oriented modernity'. Instead, 'they have remained among themselves, tied to their closest followers and locked up within their own generation' (Knapp 1999). Green core topics have rarely been an innovative influence in the politics of the coalition in Berlin. In the new citizenship laws the Greens have gained as little acceptance of their views as with their demand for a rapid withdrawal from nuclear energy. The failure to secure a new EU ordinance on old cars in 1999 was downright humiliating. Up to now the Greens have not managed to profit from the high political reputation of their Foreign Minister Joschka Fischer either. The Party's public image is weak. It has yet to find its political position – will it manage to reconcile ecology and the market economy? How does it see the balance between the individual and the State, between self-responsibility and social security? How does it intend to increase the attractiveness of Germany as an economic location in view of the growing international competition? Will the Party be able to overcome its traditional scepticism concerning technology and innovation?

The Greens are not only suffering from a saturation effect among citizens regarding the environment, but also from the fact that environmental questions were linked for so long with apocalyptic visions. The fact that there has never been a serious environmental catastrophe in Germany is clearly seen as an indication that things cannot be that bad. Against the background of new global-isation-related challenges, criticism of technology and civilisation is now considered backward-looking. Questions of location, mass unemployment and increasing social problems have displaced the environment at the top of the political agenda. Thus, the successes of the Greens and the ecology movement have been ambivalent. There is a growing interest in nuclear-power-related political and legal questions, but over the past twenty years the Greens and the anti-nuclear movement have got little closer to their goal of switching off the nuclear power stations in Germany. The radical self-realisation philosophy characteristic of the old left-wing alternative scene was not much in evidence in the red-green coalition negotiations in 1998. The Greens' participation in government at *Land* and Federal level has brought out into the open tensions between the maintenance of ideological traditions and client-orientation, in a painful confrontation with political reality. The gap between Green Party leaders and their local associations, where many activists are still attached to the ecology movement, will probably widen further, and the political identity of the party is in danger of becoming blurred. Are they a left-wing ecological reform party, are they perhaps a left of centre party, or even a citizens' rights party? Have the Greens a role as a model for the political left any longer, or do they simply reflect the loss of clear goals on the left since the collapse of socialism?

Not only the Greens, though, but also the ecology movement is facing pressure to adapt. The Greenpeace campaign against the dumping of the *Brent Spar* oil rig in the Atlantic by Shell in 1995, which aimed at symbolic confrontation (Krüger 1996), reflected a fundamental change in movement campaigning. The idealistic mass protest and demonstrations of the 1970s and 1980s are being replaced by professional, scientifically based protest actions. In the age of global non-governmental organisations (NGOs) the ecology movement seems to have reached a turning point. Whereas the protest in the 1970s and 1980s relied on direct participation and mobilisation of the masses, nowadays professional movement managers in organisations like Greenpeace direct national and international campaigns through the media. Such a hierarchically structured and highly organised 'protest industry' (Roth 1998: 57) will probably become more and more important. However, the systematic orientation of protest towards the media makes it highly dependent on trends and even commercial considerations. Alongside this, eco-political citizens' action groups will probably retain their 'watchdog' function (Ehmke 1998: 151) – though at a much reduced level from the 1970s and 1980s.

As far as the Greens are concerned, they have probably entered a new phase in their development, bidding a definitive farewell to the environment party in its original form. The party has severed its ideological roots to an extent not feared or hoped – depending on the standpoint – by most observers. Compromise, i.e., subordination to the SPD, has become the trademark of green politics. In this respect, the behaviour of Antje Radcke, spokesperson for the Party executive, is symptomatic: she chose not to stand again after the vote at the Green Party convention in June 2000 accepting the compromise with the SPD over withdrawal from atomic energy. It remains to be seen whether the voters will honour the Party's new line. If it is to be accepted, the Greens will have to offer their supporters more attractive programmes and leaders than in the past. The party convention, at which the highly-regarded *Land* politicians Renate Künast and Fritz Kuhn were elected spokespersons for the executive, could be a first step in the right direction. The Greens will need to accompany it with a convincing and credible demonstration of issue competence. Antje Radcke's withdrawal from her top office at the same time symbolises the end of an increasingly difficult partnership between the Greens and the environmental movement.

Notes

1. It should be noted, however, that since the end of the 1960s the parties, too, had witnessed membership growth – the *Sozialdemokratische Partei Deutschlands* (SPD) up to 1977, the *Freie Demokratische Partei Deutschlands* (FDP) to 1981, and the *Christlich Demokratische Union* (CDU) to 1983. See Jesse 1989: 494.
2. The figures were as follows: 1982 – 5,313, 1983 – 9,237, 1984 – 7,453. See Pfenning 1992: 32.
3. 'The Greens are a loosely-knit framework party, a relatively independent agency for left-wing libertarian ecological interests' (Raschke 1995: 17).

CHAPTER 6

The Environmental Movement and Environmental Concern in Contemporary Germany

Anja Baukloh and Jochen Roose

The close connection between environmental concern and the environmental movement seems an obvious one. People come together to express their concern about environmental problems, thus forming the movement. On the other hand, the movement seeks to raise awareness of environmental problems, thus further-ing environmental concern. However, as so often when one looks at the detail, things are not that simple. To what extent has the movement promoted environmental concern? To what extent is the movement a result of these concerns, and what exactly, for that matter, does 'environmental concern' mean?

An assessment of the overall success or failure of the German environmental movement is beyond the scope of this chapter.[1] Instead we want to limit our focus in various ways. Firstly, we will concentrate on the recent past, namely the 1990s. For these years we will look at different aspects of environmental concern and their development, distinguishing between East and West Germany, as their circumstances are quite different. We will only include inter-national comparisons to a limited extent, as such comparisons often need more contextual information than can be provided here. Secondly, our evaluation of causal relationships between movement activity and environmental concern can only remain vague. We will have to rely on plausible assumptions rather than proof for rela-tions between movement and environmental concern. In the first section we discuss the concept of environmental concern and give a short overview of the environmental movement in West and East

Germany. Thereafter we assess different aspects of this concern and consider how they relate to the movement.

The Concept of Environmental Concern

One of the few aspects that discussions of the concepts of 'environmental concern', 'environmental attitudes' and 'environmental consciousness' have in common is the observation that there is no generally recognised term (see for instance Fransson and Gärling 1999 and Dierkes and Fietkau 1988: 17ff.).[2] In the literature on the subject one finds a whole set of definitions ranging from broad usages implying general pro-environmental attitudes to narrow definitions as intentions for behaviour in everyday life (see for instance the overviews in Spada 1990: 623f. and Haan and Kuckartz 1996: 36f.). Most frequent are approaches including both aspects to a varying extent, either combined (e.g., Maloney and Ward 1973, Urban 1986) or measured separately (Dunlap and Mertig 1996, Haan and Kuckartz 1996: 37, Stern 1992: 279ff.).

On the one hand it seems useful to define environmental concern as actual preparedness for action, if not as action itself. The benefit of such a definition is that it includes a value judgement, with the subsequent costs taken into account. Studies pointing out discrepancies between measurements of environmental attitude and actual behaviour (e.g., Preisendörfer and Franzen 1996, Hines et al. 1987) support the position that environmental concern must be defined as closely connected to action. However, actual behaviour is not determined by attitudes alone, and many other factors such as available resources and constraints, peer group opinions and the perceived opportunity to have a positive effect on the environment also play a role in determining decisions about which courses of action to engage in (e.g., Tanner 1999, Wenke 1993). So if 'environmental concern' is operationalised as pro-environmental behaviour, one runs the risk of measuring the potentiality of the situation allowing for pro-environmental behaviour, rather than the attitude. On the other hand, when environmental concern is defined as a general attitude, such a definition treats a person's state of mind as an isolated factor in complex situations, whereas it is one among a series of variables explaining the resulting behaviour. Such an approach focuses on one necessary, but not sufficient condition.[3] It emerges that attitudes towards environmental protection in general and actual environmental behaviour are different though related phenomena, and follows that an assessment of the German environmental movement's influence should include both.

Another divide within the discussion of environmental concern is over personal versus political attitudes and behaviour (e.g., Waldmann 1992: 22ff., Rootes 1995, Stern 1992: 285ff.). Environmental concern has primarily been understood as an attitude manifesting itself in a preparedness for action in everyday things such as choosing private or public transport, separating waste and so on. However, for a full assessment of the environmental movement's activities the generating of political support for environmental policy cannot be ignored. The political system can have a broad impact on the environment through legislation, taxation and the like. In democratic systems like that of the Federal Republic of Germany,[4] citizens can influence policy by taking environmental concerns into account when voting, as well as through other forms of political participation. Of course, the numerous barriers that prevent or permit pro-environmental action stemming from pro-environmental attitudes in everyday life apply just as much as to political participation. Thus the distinction between everyday activity and political participation is a further dimension which should be handled separately.

Recent literature on environmental concern in its various versions has mainly focused on the question how attitudes and behaviour 'can ... be changed in a more ecological direction' (Kaiser et al. 1999: 1). For this purpose an amalgamation of the above aspects, including broader attitudes as well as actual preparedness for action and political as well as everyday issues, is necessary. A broad but differentiated approach seems appropriate for assessment of the environmental movement's role in generating new environmental concern (Jamison et al. 1990). We propose to use the term 'concern' rather than 'consciousness' or 'attitude', in order not to limit our analysis to a state of mind. This broad concept of environmental concern embraces two dimensions of dichotomy: everyday life versus political participation on the one hand, and attitude versus behaviour on the other hand. This results in four aspects of environmental concern (see Table 6.1).

Table 6.1 Environmental concern

	Attitude	Behaviour
everyday activities	*pro-environmental attitude* • support for environmental protection as a general goal	*pro-environmental behaviour* • Pro-environmental behaviour/consumption
political participation	*political support of the environment* • support of environmental concern as a political goal	*political participation on behalf of the environment* • protest (in the widest sense) on behalf of the environment • voting for green parties

These four aspects can be regarded as different ways of supporting environmental protection. Each aspect can be supported strongly or weakly by a person, and support of one aspect does not necessarily imply support of another aspect. The German environmental movement in its various organisations and groupings has sought and seeks to influence all of the aspects subsumed here under the heading of environmental concern. Before we look at each aspect of environmental concern and how non-governmental organisations (NGOs) have supported these aspects, we will give a short overview of the German environmental movement's development and current situation.

The Environmental Movement in West and East Germany

The history of the idea of protecting the environment in Germany has already been discussed by Thomas Rohkrämer in an earlier chapter in this volume. At the end of the nineteenth century, the League for the Protection of the Homeland was at the centre of a movement which called for the protection of monuments and threatened landscapes, but was not yet concerned with the environment as a whole (Christmann 1997: 39ff.). During the Third Reich, these nature protection groups were either integrated into the Nazi movement or disbanded. After the end of the Second World War some of them were re-established, however, they remained marginal, as economic and political recovery left little room for conservationist concerns.

Jürgen Hoffmann has discussed the social and political changes behind the emergence of modern environmentalism, i.e., preoccupation with the whole natural basis of human life, in the late 1960s, in West Germany alongside other Western European countries. Individual citizens' action groups protesting against construction projects in their neighbourhood such as nuclear power plants, roads and airports,[5] soon came together in a network to resist environmental destruction more generally, the Federal Association of Citizens' Action Groups for Environmental Protection, or BBU. Founded in 1972, this organisation grew into a central forum for the environmental movement in the late 1970s. However, the more traditional organisations also became active players in the movement. The Bavarian *Bund Naturschutz* (Association for the Protection of Nature, founded in 1913) played a key role in the founding of the *Bund für Umwelt und Naturschutz Deutschland* (Association for the Protection of the Environment and Nature, or BUND) in 1975,

which is now a central organisation in the environmental movement. Another major organisation in today's environmental
movement is the former *Deutscher Bund für Vogelschutz* (Association
for the Protection of Birds), founded in 1899. Its transformation
from a traditional, rather conservative nature protection organisation principally concerned with birds to a broad environmental
organisation was marked by the change of its name to
Naturschutzbund Deutschland (German Association for Nature Protection, or NABU) in 1989 (see Blühdorn 1995: 171ff.). Other major
players in the current environmental movement are the national
branch of Greenpeace (founded in 1980), the *Deutscher
Naturschutzring* (German Ring for the Protection of Nature, or DNR,
1950), an umbrella organisation for environmental groups, and to
lesser extent the national branch of the World Wide Fund for
Nature (WWF, 1963).

Though the environmental organisations were united in their
resistance to nuclear energy,[6] in other respects they have varied
considerably. While some combine practical nature conservation
with political lobbying (e.g., BUND, NABU), others offer selective
incentives for their members as service organisations (e.g.,
Verkehrsclub Deutschland, the German Transport Club), provide
scientific expertise on environmental issues (e.g., *Öko-Institut*) or
draw attention to specific problems with media stunts (e.g., Greenpeace, Robin Wood). The shifting relationship between *Bündnis
90/Die Grünen*, the German Green Party, and the environmental
movement in the 1980s and 1990s has been examined by Hoffmann above (see also Frankland and Schoonmaker 1992, Blühdorn
2000c). Especially since the Green Party became the junior partner
in government in 1998, it has been exposed to harsh criticism from
the movement (Roth 1999a). Some of the core demands
campaigned for by the movement over the last few decades have
become Government policy. However, legislation has reflected the
compromises typical of political decision-making.

In East Germany, the formation of environmental groups and an
environmental movement followed a different path.[7] As the political system of the German Democratic Republic (GDR) was
extremely hostile towards independent political activities, those
seeking to found or join environmental groups could be regarded
as engaging in high risk activism, even when political protest was
not part of a group's strategy. Until 1989, consciousness raising and,
to a greater extent, political critique were subject to severe repression. However, extreme situations, as evidenced in the highly
industrialised areas around Bitterfeld, as well as disastrous events
such as the nuclear accident at Chernobyl, functioned as focal
points for the foundation of groups, most of which were closely

associated with and protected by the Protestant Church. As early as the late 1970s, ecological problems were addressed by the East German peace movement. Criticism of the environmental situation was linked to criticism of the political system and its industrial policy (Findeis et al. 1994, Rink 2000). In 1980, in addition to the independent groups, the *Gesellschaft für Natur und Umwelt* (Society for Nature and Environment) was founded as part of the state-controlled *Kulturbund* (Association for Culture). This organisation offered some space for ecological thought and activism but remained apolitical, and had only very limited influence. In 1987, about 380 urban ecology groups were members of the *Gesellschaft für Natur und Umwelt.* At the same time, the *Stasi*, the secret service of the GDR, kept 39 ecological groups based in the Protestant Church under observation (Schönfelder 1999: 96). These groups were part of a political and social 'counter culture' in the GDR. Although the public sphere was highly controlled by the state, such groups managed to establish room for alternative thinking and discussion.

In this regard, another initiative should be mentioned: the Umweltbibliothek, a private ecological library founded in 1986 under the aegis of the Protestant *Zionskirchgemeinde* (Church of Zion parish) in Berlin. The members of the *Umweltbibliothek* had a highly critical position not only regarding environmental issues but also towards the social and political situation in the GDR in general (Kühnel and Sallmon-Metzner 1991). In November 1987, the arrest of seven members of the *Umweltbibliothek* and the confiscation of their equipment led to a number of national and international protests and helped the group gain a good deal of publicity in the West German media. As early as the next morning hundreds of citizens of the GDR went out into the streets and demonstrated against this act of state repression (Findeis et al. 1994: 27). This event counts as one of the starting points of a movement towards a peaceful revolution in the GDR.

In the last months of the GDR, environmental issues became one of the major topics of discussion. The data about the environment in the GDR had been top secret until then and the state policy was to maintain this secrecy. Meanwhile, the ecological situation had become disastrous, mainly as a result of heavy industry, which was responsible for severely polluted areas especially in the south of the GDR. The peace and environmental groups made up an important part of the citizens' movement (*Bürgerbewegung*) and their demand for radical ecological change was of outstanding importance in the autumn of 1989 (Rink 2000). During the gradual breakdown of the socialist regime the environmentalists had to take decisions about the organisational structure of the movement: while the leaders of

Die Arche (The Ark, a network of environmental groups in the GDR founded in 1987/88) as well as activists from individual church environmental groups and from the *Gesellschaft für Natur und Umwelt* opted to found a Green Party, other activists established an independent association of ecological groups, the *Grüne Liga* (Green League – see Rink 2000). In 1991, the Green Party of the GDR joined other citizens' movements to form the alliance *Bündnis 90.* They later merged with the West German Green Party as *Bündnis 90/Die Grünen.* Many new groups were founded (Rucht et al. 1997: 140ff.) among whom were local subgroups of larger West German organisations such as BUND. However, most of them were independent and loosely associated through the East German network *Grüne Liga.*

In recent years, many analysts and several authors have suggested that the movement is in crisis, in decline or even dissolving (see Hoffmann and Blühdorn in this volume, also Blühdorn 1995 and Opp 1996). This view is also frequently expressed by activists. However, systematic research has found neither a decline in the number of protest events and their radicalism (Rucht and Roose 1999), nor falling membership numbers (Rucht and Roose 2000a). This does not, of course, preclude the possibility of a decline in a different sense, such as a cultural rather than organisational institutionalisation, or an ideological decline.

Dimensions of Environmental Concern

We now take a closer look at the four dimensions of environmental concern identified in Table 1: pro-environmental attitudes and behaviour, political support for the environment and political participation on behalf of the environment. The first section on pro-environmental attitudes examines shifts in the public awareness of environmental problems. We focus especially on two topics which have been central to environmental campaigning since the 1970s: waste collection/recycling and transport. In the second section we stay with these two examples and analyse the related pro-environmental behaviour. While attitudes can easily be measured in interviews, pro-environmental behaviour is more difficult to assess. In the absence of other sources, we rely here mainly on self-reported behaviour. However, to avoid the social desirability related to the environment issue, we use specific rather than general questions. The third section gives an overview of political support for the environment in East and West Germany. With particular regard to the specific situation of German society in the 1990s, we consider the role environmental questions have played in the process of Germany's reunification as well as exploring more general survey data. Finally, in the fourth section, we look at political participation on behalf of the environment, analysing participation in environmental protest and membership figures of environmental organisations.

Pro-Environmental Attitudes

In the 1990s, pro-environmental attitudes were widespread in both East and West Germany. For example, opinion polls show that in 1997 83 percent of East Germans as well as West Germans believed that the environment was in danger and that the alarming scenarios reported in the media were realistic (Noelle-Neumann and Köcher 1997: 1055). Campaigns of the environmental movement, public debates and analyses by critical experts contributed to this public awareness. Today, environmental problems are important to an overwhelming majority and everyday behaviour seems to be at the very centre of these concerns. Asked about the measures which should be undertaken to protect the environment, in April 1993 only 23 percent of the population thought of new and better legislation, and 9 percent of more money. Fifty-six percent considered the most important factor a fundamental change in the attitude of human beings towards nature (Noelle-Neumann and Köcher 1997: 1055).

On the whole, the opinion polls give the impression that in Germany supportive attitudes towards environmental protection are widespread. There is a strong belief that fundamental changes are necessary as far as the treatment of the environment is concerned and that the state of environmental deterioration is alarming. It seems reasonable to suggest that environmental interest groups have played a significant role in influencing these attitudes. Environmental organisations have provided information and interpretation of environmental issues, thus forcing such issues onto political and media agendas.

At the same time the readiness to accept cuts in the standard of living for the sake of the environment is not very high. In an opinion poll in 1996 interviewees were asked about their degree of willingness to renounce habits and suffer material discomfort in order to protect the environment: 39 percent of West Germans and 45 percent of East Germans thought it necessary to take some form of action for the sake of the environment, but felt that other measures rather than renunciation and restriction should be found (Noelle-Neumann and Köcher 1997: 1056). In the year 2000 the number of persons unwilling to suffer material discomfort in order to protect the environment has fallen slightly to 35 percent; 55

percent were 'more or less ready' and 10 percent were 'very willing' to do so (Bundesministerium für Umwelt 2000: 36). However, there is still a clear divergence between ecological concern, the readiness to act and actual efforts.

Nevertheless, international surveys show that, for instance, when comparing results to the British population, the readiness of the West and East Germans to accept cuts in the standard of living in order to protect the environment is slightly higher: in 1993 45 percent of the West and 34 percent of the East Germans were fairly willing to accept such cuts, while only 8 percent of the West and 13.6 percent of the East Germans were very unwilling. The figures for the British population revealed 25 percent fairly willing and 20 percent very unwilling (Zentralarchiv für Empirische Sozialforschung 1999: 46).

We now take a closer look at the attitudes of the German public towards two important topics of environmental campaigning: waste collection/recycling and transport. This analysis is based on an opinion poll of 1996, carried out for the Ministry of Environment, Nature Protection and Nuclear Safety (in the following: Ministry of Environment).[8]

Waste Collection and Recycling

Regarding the attitude towards waste collection and recycling, the opinion poll points out that the German population considers the amount of waste and the environmental implications a very important issue: 49 percent of West Germans and 59 percent of East Germans are afraid of the damaging 'waste flood' and 57 percent of West Germans and 58 percent of East Germans think that the waste problem and the need for recycling are not being over-dramatised in public debate. People believe that further improvements in waste collection and recycling are possible, and 48 percent of West Germans and 60 percent of East Germans are ready to act by improving the separation and collection of waste personally. At the same time only 22 percent of West Germans and 15 percent of East Germans are ready to pay a higher fee for waste collection if this would bring about an ecologically safer waste treatment (Bundesministerium für Umwelt 1996: 32).

Taking into consideration the socio-demographic facts we can see the following situation: in West Germany, women, young people, people with higher education and voters of the Green Party are more concerned about the waste problem than men, older people, people without higher education and voters of the other political parties. In East Germany, the population in general is more sensitive towards the ecological question and there are no significant differences regarding gender, age or education. What matters in East Germany is only the voting preference: Green Party voters are more conscious of the waste problem and recycling (Bundesministerium für Umwelt 1996: 32).

Transport

The second area we want to look at is the attitude towards transport, especially towards the use of the private car. West and East Germans are united in their emotional rejection of the unnecessary use of private cars: 52 percent of West Germans and 51 percent of East Germans express anger with people who do not change their driving habits in order to save petrol, and 47 percent of West Germans and 44 percent of East Germans disapprove of people travelling by car when they could just as well take a bus, train or bicycle. On the other hand, on the cognitive level, 51 percent of West Germans and 57 percent of East Germans agree that the environmentalists are one-sided when criticising the use of cars, and only half of the population in West and East Germany consider the car one of the main causes of environmental problems in Germany. Consequently, in line with these findings, the expressed readiness to act by using the car less for environmental reasons is low: a mere 39 percent of West Germans and 31 percent of East Germans state that they try to use their car only when it is really necessary, for environmental reasons (Bundesministerium für Umwelt 1996: 50). All in all, in West and East Germany women and *Bündnis 90/Die Grünen* voters are more conscious of the environmental implications of private car use. In West Germany younger people and people with higher education also belong to this group, while in East Germany older persons are more concerned about the negative environmental impact of car use and the education level does not matter in this regard (Bundesministerium für Umwelt 1996: 52).

Pro-Environmental Behaviour

Pro-environmental attitudes do not automatically translate into pro-environmental behaviour, as everyday situations usually constrain possibilities or at least reveal the high costs of engaging in action which is not environmentally harmful. Offering alternatives which are practicable in everyday life is an important part of transforming society towards sustainability. Environmental movements have always worked towards dealing with this problem, often combining political protest and lifestyle transformation as part of their modus operandi. Analysing the overall changes in pro-environmental behaviour would go far beyond the scope of this chapter, so we restrict ourselves to the two topics already discussed regarding pro-environmental attitudes: waste collection/recycling and transport.

Waste Collection and Recycling

In the early 1990s, a new system of waste collection and recycling was introduced in Germany. As environmental groups had demanded for a long time, a system of separated collection of paper, glass and PVC was established. In some cities, organic kitchen refuse was also collected separately. The industry was required to guarantee that a significant proportion of the collected material was recycled. Recycling alone, however, does not solve the problem of the rubbish dumps generated by the consumer society, and environmental groups continue to campaign for the reduction of waste – reducing the amount of packaging, using returnable bottles, etc. – by steps that would require a fundamental change in habits. A study of attitudes and behaviour related to the new system of rubbish collection conducted in the early 1990s in two West German villages reveals that a high readiness to collect rubbish separately does not necessarily lead to other forms of pro-environmental behaviour. Most people cooperate in the state measures of separate rubbish collection, but they do nothing to reduce energy consumption or use public instead of private transport. An attempt to reduce the amount of rubbish was only made by people engaging in a high degree of reflection on pro-environmental attitudes, or under conditions where people could save money by using smaller waste containers (Schahn 1995: 255ff.).

Critics maintain that the amount of energy used to collect, transport and recycle the waste severely limits the pro-environmental effects of this new system of waste collection and recycling. Nevertheless this system has reduced the quantity of waste as well as bringing about a remarkable reduction in the volume of packaging used. In 1997 11.6 million tonnes of packaging material were used, 11/2 million tonnes less than in 1991. These figures are equivalent to 94.7 kilograms per person in 1991 and 82.3 kilograms in 1997. The amount of collected packaging material per person was 73.3 kilograms in 1997 – nearly 90 percent of all used packaging material per person (Statistisches Bundesamt 2000: 380). Even though the critical debate around the separate waste collection and recycling system has pointed out that further steps are needed to deal with the waste problem, we think that the people's willingness to participate in separate waste collection for recycling counts as a personal contribution in favour of the environment and therefore can be regarded as one example of pro-environmental behaviour.

The figures for 1996 show a high degree of participation: 71 percent of West and 74 percent of East Germans declare that they separate their waste for recycling (Bundesministerium für Umwelt 1996: 35). We should, however, consider that this data is based on an opinion poll and not on the daily observation of actual behaviour of families. The following question from the opinion poll brings us closer to real behaviour: the participants were asked how many times they threw waste into a general waste container, which otherwise could have been separated. More than 20 percent declared that they never do anything of the kind. Nearly 40 percent behaved that way occasionally, and 26 percent in the West and 29 percent in the East admitted that they sometimes threw their waste together in one container (Bundesministerium für Umwelt 1996: 35).

If we correlate the reported behaviour with socio-demographic data, the following picture emerges: in West and East Germany younger people participate less in separate waste collection than their elders. In the West, women separate their waste more often than men, and the level of education is irrelevant, while in the East people with less education tend to separate waste more than people who have attained a higher level of education. Voting preferences do not influence the participation in separate waste collection in East Germany, whereas in West Germany, *Bündnis90/Die Grünen* voters participate less in separate waste collection than CDU/CSU (i.e., Conservative) voters (Bundesministerium für Umwelt 1996: 37). We should remember that the controversial discussion of the ecological costs of the aforementioned waste recycling system could have contributed to a lower degree of participation by some people who do not regard separate waste collection and recycling a useful form of pro-environmental behaviour at all. Interestingly enough, the data of the Ministry for the Environment show that the socio-demographic groups we have found to sympathise with pro-environmental attitudes are not always identical with those who practice this behaviour more frequently. Although we have focused on separate waste collection, these results also apply to the consumption and wastage of energy.

Transport

We start our examination of the behavioural aspect of environmental concern related to transport with an overview based on car registration data. We know that the decision not to have or to use a car for environmental reasons also depends on the existence and quality of public transport, the distance to the work place and the necessity of using a private car at work. Nevertheless, we think that the number of cars per household can give an impression of the overall trend. The data of the Ministry for the Environment from 1996 shows a small West-East difference regarding the number of cars per household: in West Germany 23 percent of households do not posses a car at all (in East Germany 31 percent), 57 percent have one car (55 percent in East Germany) and 20 percent of West German households have two or more cars (14 percent in East Germany). Younger people in East and West Germany have more frequent access to a private car than older people, men more frequently than women, and the higher educated more than the less educated. Voting preference also makes a difference: voters of the Green Party own a car less frequently than the voters of other parties. However, only 5 percent of households without a car in the West and 2 percent in the East say that they do not have a car for environmental reasons (Bundesministerium für Umwelt 1996: 53).

The data on private car registration supports the impression that all the warnings of the environmental costs of private car use do not result in a change of behaviour. There has been an unbroken trend of rising private car registrations in unified Germany: in 1990 about 30 million cars were registered, in 1994 the statistics showed nearly 40 million cars on the road, and in 1999 the number had risen again to nearly 42 million (Statistisches Bundesamt 1999: 1). In East Germany, the fall of the wall in 1989 contributed substantially to the sharp increase in cars per person, which went from 50 percent of the West German level in 1988 to nearly 90 percent in the year 1997 (Statistisches Bundesamt 2000: 350f.).

In general, mobility is increasing constantly – in the working sphere as well as in leisure time. Holiday travel has increased significantly: in 1999 39 percent of the population took more than one holiday (Bundesministerium für Umwelt 2000: 53f.). The statistics show nearly 100 million people travelled in 1997 by aeroplane – 60 percent more than in 1991 (Statistisches Bundesamt 2000: 355). Use of the train as a means of transport is relatively low: in 1999 only 15 percent of the population used the train often or very often; 18 percent travelled sometimes by train, 30 percent occasionally and 37 percent never. At the same time, the position of the private car remains dominant; for shopping, going to work or travelling during the holidays (Bundesministerium für Umwelt 2000: 53f.).

This general increase of mobility is also reflected in the data on energy consumption. There has been a continuous increase of energy consumption for traffic (all traffic, not only private) from about 17 percent of total energy consumption in the year 1950 to nearly 28 percent in the year 1997. At the same time energy consumption in the industrial sector has fallen consistently since increasing in the 1950s: in the year 1950 the figure was nearly 46 percent, but already by 1970 industrial energy consumption had fallen to less than 40 percent. It fell further in 1980 to 34 percent, and in 1997 to just over 26 percent of the total energy consumption in Germany (Statistisches Bundesamt 2000: 371).

While the environmental movement seems to have had an impact on the attitude towards the environment (see above under 'Dimensions of environmental concern: Pro-environmental attitudes'), the effect on pro-environmental behaviour seems very small. The study of the German Ministry for the Environment indicates a negative correlation between a pro-environmental attitude and pro-environmental behaviour. In practice in daily life the most concerned seem to be the least pro-environmentally active, except for private car use as evidenced by Green Party voters, who are less likely to own a car than the voters of other parties. In general, personal economic resources seem to play a prominent role in pro-environmental behaviour. Limitations in economic resources seem to contribute much more to pro-environmental behaviour than pro-environmental attitudes do.

Nevertheless the very core of environmentalists do not just pay lip service to environmental problems. A survey among the members of BUND reveals that environmental activists not only show above average pro-environmental attitudes, but also orient their personal behaviour along environmental lines. In the sphere of alternative consumption and separate waste collection, BUND members show an extremely high degree of pro-environmental behaviour, while in that of mobility the differences from the rest of the population are less spectacular, but nevertheless significant. BUND members are less likely to own a car, use the train significantly more often, and fly less frequently than the rest of the population (Bodenstein et al. 1998). We can assume that these findings are also valid for the members of other environmental groups.

Political Support for the Environment

In the autumn of 1989, when the GDR government was in crisis, environmental issues spurred several protests and environmental concerns were part of the political programmes of the East German citizens' movement (see above under 'The environmental movement in West and East Germany'). A reflection of the importance attached to environmental policy can be seen in the constitutional draft for the GDR the Round Table – made up of representatives of the old regime and members of the citizens' movement – prepared in the winter of 1989/90. In addition to social rights, the Round Table wished to include environmental protection in the constitution as an objective of the state. After the first free election in March 1990 the newly elected Parliament of the GDR did not accept the constitutional draft of the Round Table, and with the unification of Germany the West German *Grundgesetz* (Basic Law) became the constitution for East and West Germany. Two years later, the idea of including environmental protection in the constitution had a second chance in the *Gemeinsame Verfassungskommission* which was set up by the parliament in May 1992 to discuss amendments to the constitution.[9] The public debate on constitutional reform following German unification was highly controversial. While some demanded that the West German Basic Law should remain unchanged, others argued that the expectations and experiences of East Germans should be reflected in constitutional reform. They referred especially to social rights (e.g., in employment, housing, social welfare, etc.) and the implementation of direct democratic elements in the constitution. In the end, only one demand of the East German *Bürgerrechtsbewegung* (citizens' rights movement) found its way into the constitution: the protection of the environment as an objective of the German state (Article 20a *Grundgesetz*).[10] This East German demand was supported by a West German cross-party consensus that the necessity of environmental protection should be introduced into the German constitution (Batt 1996: 120).

This constitutional amendment corresponded with public opinion on environmental protection as a political goal. In the year 2000 efficient protection of the environment is considered a very important political objective by 53 percent and an important political objective by 41 percent of citizens. Only 6 percent consider protection of the environment less important and none consider it not important at all (Bundesministerium für Umwelt 2000: 17). When we look at the development over time of the percentage of citizens who consider protection of the environment a very important political objective, we see that the numbers develop with ups and downs between 40 percent and nearly 60 percent from 1993 to 1999 (Bundesministerium für Umwelt 2000: 18).

Meanwhile, the population is quite sceptical about the performance of the State in the sphere of environmental protection. In the year 2000 half of the population (53 percent) do not think that the existing laws protecting the environment are adequate. Thirty-four percent consider them adequate and 13 percent are undecided. In comparison over time we see that approval of environmental legislation has increased consistently from 1991 (23 percent) to 1998 (37 percent), and then fallen back slightly in the year 2000 to 34 percent of the population who consider the existing legislation adequate (Bundesministerium für Umwelt 2000: 33). The decrease of approval in the year 2000 may perhaps be explained by unfulfilled expectations relating to the new Social Democrat and Green

Government. The results of efforts undertaken to protect the environment are also judged quite critically. As empirical data from 1997 indicate, only 34 percent speak of 'big steps' that have been made, 49 percent consider efforts undertaken 'small steps' in the right direction, and 13 percent do not see any progress at all (Noelle-Neumann and Köcher 1997: 1065).

Environmental groups and citizens' initiatives are considered especially capable of contributing to the solution of problems concerning the protection of the environment, followed by the State agencies for environmental protection and the Green Party. People are more or less confident in the environmental problem-solving capacity of the Social Democratic party SPD, the trade unions and the churches. At the bottom of the ranking we find the Conservative Party CDU/CSU, the Liberal Party FDP, the Socialist Party PDS and the industrial sector. This ranking has remained nearly unchanged between 1998 and 2000 (Bundesministerium für Umwelt 2000: 35). Correlating the level of trust in the environmental problem-solving capability of an organisation with individual interest in politics, we find that the more a person is interested in politics, the higher his/her trust is in the capability of environmental groups, citizens' initiatives and the political parties to solve environmental problems (Bundesministerium für Umwelt 2000: 34).

Political Participation on Behalf of the Environment

The environmental movement has primarily engaged with the political system in two ways: on the one hand the Green Party seeks to win votes at elections; on the other hand environmental groups mobilise support for a wide range of protest activities.

In their early years, as Jürgen Hoffmann has pointed out above, *Die Grünen* were closely connected and identified with the environmental and peace movements. However, after their electoral defeat in 1990, when the West German Green Party failed to pass the 5 percent hurdle, the party leaders began to professionalise and explicitly 'normalise' the party. Since *Bündnis 90/Die Grünen*[11] formed a coalition government with the Social Democrats, the gap has widened between the movement and the Green Party, especially its national leadership.[12] The Greens have been harshly criticised by movement organisations for their role in government, and in the *Länder* the electoral results for the Green Party have been marked by an overall decline (Blühdorn 2000c).

However, voting decisions do not reflect environmental concern directly. The decline in the Greens' electoral success can even be seen as a criticism of their failure to deliver in key areas of environmental policy. Though opinion polls show that in recent years economic issues have become more important for voting decisions than environmental concern, the internal problems of *Bündnis 90/Die Grünen* have at least partly accounted for their problems in recent elections.

In contrast, environmental protest actions are more closely related to environmental concerns. Since about 1970, environmental protest has become widespread in West Germany, starting moderately in the early 1970s and growing in the 1980s (Rucht and Roose 1999). The period we wish to analyse more closely[13] is overshadowed by German unification. Figure 1 shows the economic consequences of this event dominating the political agenda:[14]

Figure 6.1 Environmental protest events in Germany 1988–1997

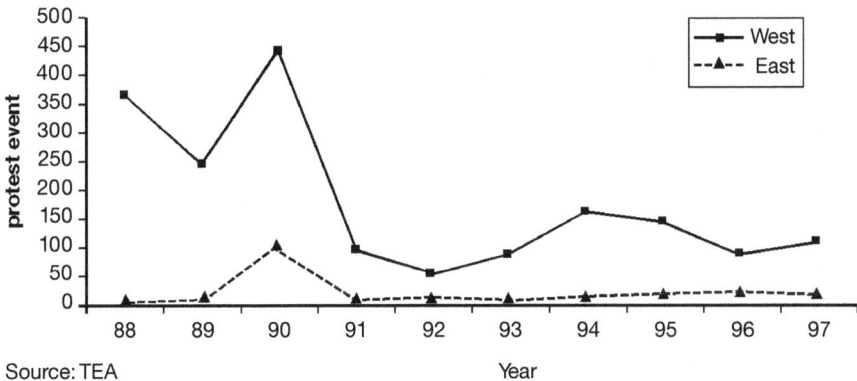

Source: TEA

Besides the relatively constant number of protest events over the years in both parts of Germany since the early 1990s, the major importance of environmental issues in the East during 1990 is worth noting. The average number of participants in environmental protests in the West is greater than the corresponding number in the East for most years. Overall there seems to be a stable number of both protest events and participants over the years, a success which can be directly attributed to the activities of environmental NGOs. The number of protest events in the East is far lower than in the West; however, one should bear in mind that protest actions in Berlin (16.7 percent of the total) are excluded. A significant percentage of these events were probably organised by East rather than West Berlin groups.

International comparisons of the frequency of protest events based on newspaper reports are problematic. In their study, Kriesi et al. (1995: 22) suggest a high level of environmental protest in Germany and Switzerland and less in France and the Netherlands.[15]

Preliminary results from the TEA-project suggest that the level of reported environmental protest in Germany and Britain is very high compared to the other countries studied.[16] However, to assess the comparability of the newspapers would require an internationally standardised source. This is obviously not available, so caution must be taken with comparisons.

Another form of pro-environmental political engagement is the support of environmental NGOs themselves. These organisations not only organise protest, but also put pressure on politicians in many other ways, e.g., by providing counter-expertise, lobbying, etc. A high level of membership increases an organisation's influence in the policy-making process. The membership of German environmental NGOs has risen during the last decade. This trend can be seen in the large environmental organisations, i.e., BUND, NABU, WWF and Greenpeace (Rucht and Roose 1999), and a closer investigation of local groups in Berlin for the years 1988 to 1997 showed that these smaller groups have also increased their membership (Rucht and Roose 2000a).

Systematic data on membership differentiating between East and West is seldom available. It is clear though that the membership of BUND, NABU and Greenpeace is much lower in East than in West Germany (Rink 2000). The NABU's membership expressed as a percentage of the population varies for the West German *Länder* between 2.3 percent and 12.3 percent, while the respective percentages for the East German *Länder* range from 0.9 percent to 1.7 percent (source: NABU annual report 1998). Other independent groups who have deliberately resisted joining the West German-dominated organisations cannot make up for the sizeable difference. The environmental movement's organisational structure is much weaker in the East than its counterpart in the West. These impressions are further supported by the results of the survey on environmental consciousness (Bundesministerium für Umwelt 2000: 71). While in West Germany 22 percent of the population would consider voluntary work in the field of environmental protection, the figure in the East is only 16 percent.[17]

As with international comparisons of protest intensity, comparison of membership numbers in environmental movements is problematic.[18] For an assessment of the movement's membership it is crucial to define which organisations are regarded as part of the movement. This, however, depends on the history of the movement.[19] A broad assessment of membership of environmental NGOs suggests that the German movement is fairly strong compared to other European countries. Overall the membership levels show that there is a growing commitment to environmental NGOs in Germany as a whole, in the West more than the East.

The number of protest events decreased after unification in 1990, but has recovered to some extent thereafter. The environmental NGOs have been able to maintain a considerable level of political participation on behalf of the environment in both their own membership and environmental protest actions. However, one should bear in mind that only a minority of the population participates actively.

Conclusion

Environmental concern plays a significant role in German society, as our analysis has shown. Yet this has happened to a different extent in each of the four areas we discuss in this article. Pro-environmental attitudes are widespread, and environmental problems are constantly high on the agenda. There are no significant differences between West and East Germans in this respect: in the opinion polls West Germans express only a slightly higher level of pro-environmental attitudes. Though NGOs are obviously not the sole source of awareness of environmental problems, they have played a key role in bringing them to the attention of the public. Thus consciousness raising can plausibly be regarded as a major success of the environmental NGOs. The findings on pro-environmental behaviour show a completely different picture: though a majority think there must be changes in everyday behaviour, individual behaviour has only changed to a small extent. In areas where it is particularly easy to become active, such as waste separation provided directly on the doorstep, people do so. Accepting sacrifices like higher costs or giving up one's car remains a rare occurrence – in West as well as East Germany. A fundamental change in behaviour has not taken place and the environmental NGOs' influence seems to be small. Economic resources seem to be the key variable influencing people's behaviour. Only a core group of environmentalists are changing their behaviour and accepting the resulting sacrifices.

Nevertheless for a huge majority of the German people environmental protection is an important political goal. Whereas the performance of the Government is evaluated sceptically by the German people, environmental groups and citizens' initiatives are highly valued. Their contribution is believed to be crucial in solving environmental problems. However, this overall confidence in environmental groups and citizens' initiatives does not lead to a huge personal engagement in these organisations. In both East and West only a minority participates in protest or in environmental NGOs. Nevertheless, in West Germany the NGOs have gained a stable level of support. Membership numbers and budgets have grown and environmental protest is widespread. In East Germany the large membership organisations have not found wide support and small local groups seem weaker. The long history of the West German movement and the relative youth of the movement in the East are reflected in these figures.

The conclusion that can be drawn from these results is twofold. It seems that the German environmental movement has had a considerable impact on public opinion with regard to both general pro-environmental attitudes and political support for environmental protection. At the same time, people's behaviour has not fundamentally changed, and only a small minority orient their behaviour along environmental principles. It remains a matter for debate whether this state of affairs should be regarded as a success or a failure of the German environmental movement.

Notes

1. The problems surrounding the analysis of the impact of environmental movements are discussed in Rucht 1999b.
2. While discussion in German refers only to 'Umweltbewußtsein' (environmental consciousness), in British and American research all three terms, i.e., attitude, consciousness and concern, are used.
3. Proposals for a broader theoretical framework may be found in the recent overviews by Lehmann (1999) and Fransson and Gärling (1999).
4. We focus on the last ten years, when a democratic system has been installed in East as well as West Germany. However, in the course of our analysis we will have to keep in mind that the political situation in the East has remained influenced by the GDR's political system.
5. For literature on the West German environmental movement see Brand 1999b, Rucht and Roose 1999 and Rucht 1994. For environmental organisations see Blühdorn 1995, Cornelsen 1991 and Rucht 1991.
6. Blühdorn calls it 'the common denominator of all movement organisations and initiatives' (1995: 172).
7. There has been little literature on the East German environmental movement up to now. Nevertheless, the environmental movement is mentioned in works on the East German *Bürgerbewegung* (citizens' movement) as a whole. The following remarks are mainly based on Rink 2000.
8. This opinion poll was conducted by the institute GFM-GETAS. 2,307 citizens were questioned (1,095 from West Germany and 1,212 from East Germany) and their responses analysed by Andreas Diekmann and Peter Preisendörfer of the Universities of Bern and Rostock (Bundesministerium für Umwelt 1996). The most recent opinion poll for the Ministry of Environment, carried out in the year 2000, was conducted by the EMNID-Institute with 2,018 citizens. The analysis was undertaken by Udo Kuckartz of the University of Marburg (Bundesministerium für Umwelt 2000). This opinion poll, however, excludes waste collection and recycling, and the questions on transport are mainly concerned with the effects of governmental measures. We therefore base our comments in the following on the 1996 opinion poll, and include some comparisons with the year 2000 concerning transport only in the next section on pro-environmental behaviour.

9. The *Gemeinsame Verfassungskommission* consisted of members from both the parliament and the *Bundesrat* (the Upper House, an institutional body of the German *Länder*). The commission presented its final draft in May 1993.

10. Article 20a: 'The state protects the natural environment, in responsibility for future generations, within the bounds of the constitution, through legislation and regarding law through the executive and the jurisdiction.' (translated by the authors)

11. The West and East German Green Parties merged to form *Bündnis 90/Die Grünen* in 1991. See Section 'The Environmental Movement in West and East Germany'.

12. An analysis of protest events shows that local and regional groups of *Bündnis 90/Die Grünen* continue to be involved in a considerable number of environmental protest activities (Rucht and Roose 2000b).

13. Our data are based on a protest event analysis of reporting in the German left wing newspaper *die tageszeitung* (*taz*). This data set is part of the project 'Transformation of Environmental Activism (TEA)', which compares environmental movements in seven countries between 1988 and 1997. For details see Rootes 1999a.

14. For our analysis of protest events in East and West Germany only events with an indication of locality are included. Furthermore from 1990 on protest in Berlin cannot be defined as East or West German protest and is excluded. Thus 17.9 percent of the total protest events from 1988 until 1997 (N=2,470) are excluded.

15. Both the TEA project and Kriesi et al.'s study include protest against nuclear energy.

16. The other countries are Italy, Spain, Greece, France and Sweden. See Rootes 1999b.

17. The study also reveals that the overall proportion of the population taking part in voluntary work in various areas including the environment is much lower in the East (4%) than in the West (9%). Data on actual activism in the environmental sector alone are, however, not available.

18. For a discussion see Rootes and Miller 2000.

19. For example in Britain and the Netherlands, large organisations (the National Trust and *Natuurmonumentum*) combine the preservation of landscapes with that of the built environment, and offer their members free entry to the organisations' properties. These organisations have a huge membership in both countries and there is no functional equivalent in Germany. However, in their respective countries they are active in the environmental movements.

CHAPTER 7

Green Futures? A Future for the Greens?

Ingolfur Blühdorn

Three decades after ecological issues first appeared on the political agenda of Germany and most other industrialised countries, social movement researchers and environmental activists have sought to assess the achievements and failures of *green* politics[1] so far (e.g., Blühdorn 1995, 2000a, Diekmann and Jaeger 1996, Rucht 1996, Jacobs 1997, Ehmke 1998, Klein et al. 1999, Rootes 1999c, Rucht and Roose 1999, 2000a). On the one hand they can point to an impressive record of success in terms of generating public environmental awareness, putting environmental concerns on a strong institutional footing and initiating actual policy changes. These achievements have given rise to optimistic hopes that we might, after all, be heading towards a green future, and that the century we have only just entered might become the 'century of the environment' (von Weizsäcker 1999). On the other hand, however, there is very little evidence that the great ecological U-turn which political ecologists have demanded ever since the 1970s has really been taken or is about to be taken. On the contrary, throughout the 1990s new economic, political and cultural constellations have refocused public attention on issues of material security and economic growth, and have fundamentally changed the way in which ecological issues are being framed and tackled.

This chapter focuses specifically on the developments since the early 1990s and aims to explore the future prospects of the eco-movement and political ecology. Using the German experience as its prime point of reference, it argues that at the beginning of the so-called *century of the environment* advanced modern societies are not moving towards the *realisation* of the green ideals and *ecologist*[2] visions that once provided the intellectual core and stimulus for the eco-movement, but instead towards their *abandonment*. The central hypothesis is that although ecologist thinking has undoubtedly had an impact on the way contemporary societies shape their political debates and policies, there will be neither a green future in the ecologist sense, nor a future for ecologist greens. This argument will be developed in three main steps. The first section focuses on the recent stagnation of German eco-politics. The second section explores the potential for its revitalisation, and the last section discusses the irreversible break-up of the eco-ideological package-deal. The transformations this chapter tries to capture are not exclusive to Germany. *Mutatis mutandis*, they arguably affect late modern societies in general.

Eco-Political Stagnation

National Environmental Politics

Shortly after Tony Blair's New Labour government had put an end to eighteen years of Conservative rule in Britain, Michael Jacobs noted that 'politically the environment is marginal', that it 'fails to excite conflict between the major political parties', and that in the election campaign the environment 'simply never made it onto the agenda' (1997: 3-5). In the same publication Neil Carter added that the 'major political parties, after taking a crash course in environmentally friendly rhetoric, have since returned to *politics as usual* – the economy, taxation, health, education, law and order' (1997: 192). Of course, Britain has never been known as a particularly eco-friendly EU-country. However, these observations on Britain – which are no less appropriate today than they were at the time – seem equally applicable to most other European countries – including Germany, once at the forefront of European eco-politics. Looking back on the Kohl era, Weidner and Jänicke note that the Federal Republic has built up considerable *eco-political capacities*,[3] but that in recent years, due to 'insufficient will and skill', the 'existing potential for action is not being realised'. In 'certain areas' they even identify 'a loss of eco-political capacities' (1998: 202). Karl-Werner Brand notes that in particular since the beginning of the 1990s, there is 'stagnation' and a 'crisis of environmental politics' (1999b, 1999a). Although the level of general environmental awareness and concern has remained high (Preisendörfer 1998, Baukloh and Roose in this volume), unification clearly changed the list of public concerns and political priorities. Undeniably, there is – most notably in the new *Länder* – a good record of eco-political achievements. However, these improvements are primarily related to 'environmental problems of high visibility and potential for politicisation' (Weidner and Jänicke 1998: 221). Crucial developments like the exhaustion and contamination of fresh (ground) water, the loss of biodiversity, the expansion of built-up areas, rapidly increasing car and air travel, summer smog, etc. could not be stopped or reversed. For these issues there is no straight-forward technological solution, and – perhaps more importantly – there is no immediate economic gain to be expected from addressing them.

In the campaign for the German Federal elections in October 1998, environmental issues – just as Jacobs noted for Britain – did not play any significant role (Semetko and Schoenbach 1999: 81). Topics like the phasing out of nuclear energy or plans for an ecological tax reform were occasionally mentioned, but the prospective new Chancellor, Gerhard Schröder, was keen to present himself as friendly to the business world 'Genosse der Bosse' and sympathetic to middle class motorist enthusiasm 'Kanzler aller Autos'. He was therefore restricted to a very moderate and non-conflictual brand of environmentalism. The debate on petrol prices earlier in that year[4] had made it unmistakably clear that environmental issues were not a public priority. Nevertheless, the formation of the first ever coalition government in Germany involving the Greens gave rise to high hopes amongst environmentalists for significant ecological progress. By the middle of 2000, however, these hopes had largely evaporated. Despite some modest progress in terms of eco-taxes[5] and the nuclear issue,[6] the major environmental organisations in Germany were complaining, unanimously, that the excellent opportunity for bringing new momentum into environmental politics had been missed. In its *Environmental Report 2000* the Federal Government's *Sachverständigenrat für Umweltfragen* (Council of Experts for Environmental Issues) notes that the country, which in terms of environmental politics used to be an 'international forerunner', is 'today amongst the latecomers' (2000: Section 1). Whilst a number of other European countries have long begun to implement plans and strategies of ecological modernisation, Germany has hardly started to develop a national strategy for the achievement of environmental sustainability. In line with Carter's comments on Britain, Karl-Werner Brand notes that in Germany, too, 'traditional arguments of growth and costs are once again pushing the discourse of ecological modernisation into the background' (1999a: 245). Some time before the 1998 elections Detlef Jahn predicted that 'the environment does not look like becoming a major issue again' (1997: 181). Up to the present there have been 'few signs of further progress', and 'no major breakthroughs can be expected' (Rucht and Roose 1999: 59).

The Eco-Movement

Of course, the prediction that environmentalism and political *greenery* are no more than a passing fad is as old as the eco-movement itself. It is therefore appropriate to handle such prophecies with caution and ask for their analytical foundation. Already at a very early stage environmental sociologists have suggested that the ecological issue is subject to *issue attention cycles* (Downs 1972), which means that after a phase of increased information and mobilisation, the media and the general public will gradually lose interest, with the actual impact of the wave of public concern on concrete politics remaining marginal. The ups and downs of issues like nuclear power, acid rain, dying forests, climatic change or genetic modification have provided ample evidence that the environmental debate is indeed determined by issue attention cycles. However, it is as yet unclear to what extent – or whether at all – this also applies to the environmental issue as a whole. Looking at the organisational structure of the environmental movement in Germany, one might come to the conclusion that at the beginning of the twenty-first century at least the environmental movement – if not the eco-politics of the Government – is in excellent shape and set for major progress. Environmental organisations now command significant resources in terms of membership, budget, staff, scientific expertise, and formal and informal channels of influence. Table 7.1 shows that in recent years the major environmental NGOs have enjoyed a constant increase in both the number of supporters and their annual budget.

Table 7.1 Resources of major environmental NGOs in Germany

	1994	1995	1996	1997	1998	1999
GREENPEACE incomings[7] regular donors	67.2 507.000	69.1 517.500	66.9 517.000	65.1 520.500	68.7 531.500	65.4 510.000
WWF incomings regular donors	27.5 146.000	28.4 160.000	28.7 180.000	27.4 185.000	34.7 199.000	39.0 225.000
NABU incomings members	15.5 185.100	17.5 195.000	18.2 210.000	18.8 210.000	20.8 225.000	22.5 241.000
BUND incomings members & reg. donors	16.7 225.000	19.9 253.200	20.5 285.700	22.5 316.500	27.0 345.900	30.4 364.300

(source: annual reports of individual organisations)

Dismissing concerns that the eco-movement might be running out of steam, Rucht and Roose therefore insist that in contrast to the government's official environmental politics, 'the movement is not in decline' (1999: 59). More recently, they have reconfirmed the view that there is 'neither decline nor sclerosis' (2000a). This assessment is certainly in line with a general shift from the earlier eco-pessimistic rhetoric to a much more optimistic and forward-looking perspective in both recent academic literature and public debate (e.g., Maxeiner and Miersch 1996, Bohnke 1997, Weizsäcker 1999, Fischer et al. 2000). However, focusing the perspective too narrowly on the impressive growth of the major organisations may easily generate a false impression. Firstly, such an approach does not take into account that the rising figures for membership and organisational income are, to a significant extent, the result of increasingly sophisticated methods of targeting membership and donation recruitment rather than an increasing public interest in and commitment to environmental issues (see below). Secondly, such an approach might distract attention from the negative implications of the *institutionalisation* and *professionalisation* of the eco-movement (e.g., Roth 1994, Blühdorn 1995, Opp 1996, Brand 1999a, 1999b, Rootes 1999c). In a quite Weberian sense, these developments imply that processes of *rationalisation* and *bureaucratisation* gradually formalise the environmental debate, hollowing out its substantive content (see section below) and leading towards a state of bureau-cratic sclerosis and decline. Thirdly, an excessively narrow focus on the growth of institutionalised resources fails to capture the funda-mental changes which elsewhere I have tried to describe with the concepts of the *temporalisation, individualisation* and *aestheticisation* of the eco-movement (Blühdorn 2000a). What these concepts mean is that waves of protest and mobilisation (issue-attention cycles) are becoming ever more media-dependent and short-lived; that reflect-ing the interests of both professional organisations and the wider populace, the quality of environmental engagement has clearly changed from patterns of collective involvement and direct action towards private credit card activism and representational protest; and that the self-sacrificing and altruistic ethos typical of earlier patterns of eco-activism has given way to an emphasis on fun, self-construction and self-experience.

Not coincidentally, the annual *Love Parade* in Berlin – officially claiming the status of a political demonstration rally – has, through-out the 1990s, attracted hundreds of thousands of participants (in recent years in excess of a million), whilst the Federal Association of Citizens' Initiatives for the Environment (BBU), for example – for a long time the backbone of the German eco-movement – has been struggling for survival. Fully professionalised and well-resourced organisations like Greenpeace and the BUND, on the other hand, face a different kind of problem: they are uncertain how – and for which concrete political goals – their impressive resources should best be used. These organisations are confronted with a major ideological crisis and disorientation. As the ecological certainties (simplifications) of the 1970s and 1980s are becoming increasingly questionable, the main challenge today is the 'general disillusion and insecurity about what can be done' (Rucht and Roose 1999: 76). It is therefore fully appropriate if Rucht and Roose, in contradiction of their optimistic comments quoted above, also believe that the German environmental movement has arrived 'at a crossroads', facing 'a stalemate which one could associate with a mid-life crisis' (1999: 74). Arguing along very similar lines, Karl-Werner Brand speaks of 'severe identity problems' for the eco-movement (1999b: 35). Obviously, terms like 'cross-roads', 'mid-life crisis' and 'identity problems' have been carefully chosen to imply that there is still scope for re-orientation, regeneration and re-radicalisation. However, as I will argue below, the framework parameters for any such revitalisation of the eco-movement do not look particularly favourable. Before these parameters are explored in more detail, a brief look at the German Green Party seems in order.

Bündnis 90/Die Grünen

Already in 1994, just before *Bündnis 90/Die Grünen* re-entered the German *Bundestag*, thus putting an end to the unexpected – and then undeserved – four year period during which the Western Greens had been consigned to extra-parliamentary opposition, Anna Bramwell had diagnosed *The Fading of the Greens* and the 'decline of environmental politics in the West' (Bramwell 1994). At the time, the resurrection of the Greens as a parliamentary party seemed to provide sufficient evidence that Bramwell's hypothesis was wrong. In October 1994 the Greens secured 7.3 percent of the votes, relegating the liberal democratic FDP (6.9 percent) to position four. Four years later they then actually replaced the FDP as the junior partner in the new coalition government (Greens: 6.7 percent; FDP: 6.2 percent). Being involved in the Federal Government and securing three ministerial portfolios[8] seemed the ultimate victory for political ecology. However, compared to their 1994 results *Bündnis 90/Die Grünen* had actually lost votes. Furthermore, the social composition of the Green clientele had changed significantly. Although in terms of the membership base the second half of the 1990s meant consolidation and even a slight increase of support for the Greens (Table 7.2), a disproportionately large number of young voters (age group 18-24) – in both the old as well as the new *Länder* – had turned away from the Greens (minus 4 percent in the West and minus 2 percent in the East).

Table 7.2 Membership of *Bündnis 90/Die Grünen*

	1995	1996	1997	1998	1999
old Länder	43,583	45,281	46,031	48,562	47,801
new Länder	2,827	2,753	2,949	3,250	3,096
total	46,410	48,034	48,980	51,812	50,897

(source: *Bündnis 90/Die Grünen*)

As in Britain, where the Green Party has 'lost teenage support from six per cent in 1994 to a mere two per cent in 1998' (Jowell et al. 1999: 30), the decrease in support for *Bündnis 90/Die Grünen* amongst young people can only be described as dramatic (see Table 7.3). The losses were not restricted to the first-time voters either, for in the age group from 25 to 34 they had also lost 3 percent in the West, (plus 2 percent in the East; Gibowski 1999: 24f.; also see Hoffmann in this volume). These shifts have become a standard pattern in the various elections held in recent years. The German Greens, unable to tune into the Zeitgeist of post-unification youth, seem to remain tied to the milieu of the (nostalgic) post-1968 protest generations.

Table 7.3 Party identification amongst young people (age 15–24, in percent)

	1996	1999
SPD	20.0	21.2
CDU/CSU	15.4	21.7
Bündnis 90/Die Grünen	21.6	11.1
PDS	2.8	2.9
FDP	2.1	2.0
Republikaner	2.4	1.8
other	2.5	3.5
none	32.7	35.9

(source: Fischer et al. 2000: 265)

Once in government, *Bündnis 90/Die Grünen* found it extremely difficult to establish their own political profile and realise any of their central demands. With regard to eco-taxes and the phasing out of nuclear energy, it was actually the SPD which claimed political ownership[9] of these issues, and the results of their initiative were political compromises on the basis of the lowest common denominator. The high speed rail connection between Hamburg and Berlin (*Transrapid*) was stopped for economic rather than ecological reasons. Despite the strong resistance of most Greens, German troops did also participate in the Kosovo war. During this war, the Green Foreign Minister Joschka Fischer, who had actually supported the German involvement in Kosovo, emerged as the most popular German politician, yet he was keen to retain a clear distance from his own party, and in particular he sought to keep clear of any ecological issues. Amongst the German public, the green *Realissimo* was hardly associated with eco-politics at all. The former *Fundi* and now Environment Minister Jürgen Trittin, on the other hand, did try to push ecological issues – and quickly emerged as one of the most unpopular politicians in Germany. At least in the earlier part of his term as Minister for the Environment, Trittin was undoubtedly a credible and vociferous proponent of green grass-roots demands, yet he was unable to communicate his policies to the wider electorate. More recently, when Trittin had to support and manage the shipment of nuclear wastes to Gorleben (first half of 2001), the support and confidence of movement activists and major NGOs quickly melted away.

Table 7.4 shows that up to the present *Bündnis 90/Die Grünen* have not managed to convert the publicity bonus they have as a partner in government into electoral capital. Although Fischer retained his high ratings in the political popularity charts and provided major stimuli for the debate on European integration, his party has not managed to give any prominence to ecological issues. Even BSE and Foot and Mouth Desease, which triggered major waves of concern in the early parts of 2001, and very powerful demands for a radical ecological restructuring of German and European agribusiness, in the end remained a short episode with no significant consequences. Against the background of high unemployment figures and the debate on Germany's competitiveness as an industrial location in an increasingly globalised economy (*Standortdebatte*), environmental issues are simply not in demand. Whilst Schröder has spent most of his first term in office trying to restyle his party and modernise social democracy following the model of Tony Blair's New Labour,[10] *Bündnis 90/Die Grünen* have been paralysed by internal tensions and the attempt to resolve their major identity crisis. Looking ahead to the Federal elections in 2002, the prospects for the Greens are rather gloomy. However, even in the short term their political survival seems under serious threat.

Table 7.4 *Bündnis 90/Die Grünen* in *Länder* elections since January 1998:

	Federal Elections 10/1998 by Federal State	Länder elections since 1/1998 (compared to previous)
Baden-Wurttemberg	9.2%	3/01 7.7 (-4.4%)
Bavaria	5.9%	9/98 5.7% (-0.4%)
Berlin	11.3%	10/99 9.9% (-3.3%)
Bremen	11.3%	6/99 9.2% (-3.9%)
Hamburg	10.8%	
Hesse	8.2%	2/99 7.2% (-4.0%)
Lower Saxony	5.9%	3/98 7.0% (-0.4%)
North Rhine-Westphalia	6.9%	5/00 7.1% (-2.9%)
Rhineland-Palatinate	6.1%	3/01 5.2 (-1,7%)
Saarland	5.5%	9/99 3.2% (-2.3%)
Schleswig-Holstein	6.5%	2/00 6.2% (-1.9%)
Brandenburg	3.6%	9/99 1.9% (-1.0%)
Mecklenburg-West Pomerania	2.9%	9/98 2.7% (-1.0%)
Saxony	4.4%	9/99 2.6% (-1.5%)
Saxony-Anhalt	3.3%	4/98 3.2% (-1.9%)
Thuringia	3.9%	9/99 1.9% (-2.6%)

A Revitalisation of Eco-Politics?

New Radicalisation?

In the recent literature, there is some talk of a re-radicalisation of the eco-movement, both in Germany and in other major industrial countries. Anti-road protests in Britain, anti-*Castor*[11] protests in Germany, direct action against outdoor experiments with genetically modified food crops, radical protests at World Trade Organisation and G8 meetings, etc., all seem to indicate that a rejuvenation and re-radicalisation of the environmental movement is at least conceivable (e.g., Doherty 1999, Rucht and Roose 1999, 2000a, Brand 1999b). It has been suggested that 'the very fact that "social movements tend to be absorbed by the system" provides a motive for the radicalisation of other groups' (Rucht and Roose 1999: 71). Christopher Rootes even speaks of an '*iron law* of democracy which dictates that cumbersome bureaucratic organizations', both outside and inside the eco-movement, 'are sooner or later – or perennially – challenged by new, uninstitutionalized forms of collective action' (1997: 329, my emphasis). Arguing from the point of ideological rather than organisational analysis, it has also been suggested that 'the concept of *sustainable development* provided the basis for strategic reorientation and a new opportunity for public mobilisation' (Brand 1999b: 42).

It is undoubtedly correct that in recent years, particularly the nuclear issue has triggered some protest and civil disobedience in Germany. However, it is premature to take such incidents as indicators of a new wave of large-scale mobilisation. Firstly, this radicalisation of protests within small segments of the eco-movement does not reverse the larger trend towards de-radicalisation, de-ideologisation and problem-specific pragmatic approaches. Secondly, it is also true that 'alongside the new social movements, the system of internal security' has been significantly expanded, and that with the security forces in many cases easily outnumbering the actual protesters, 'the policing of the protests often replaces the discussion of the actual issues' (Roth 1999b: 60). Thirdly, neither the science-oriented and abstract concept of sustainability nor the radical strategies of marginal groups of eco-activists appeal to the wider public. As means of providing the eco-movement with a new mobilising consensus and ideological orientation they are equally inappropriate. For these reasons, belief in a new phase of mass mobilisation seems unwarranted. As a matter of fact, there is evidence to suggest that in the foreseeable future a further de-radicalisation of environmental protests and politics is the much more probable scenario.

Shifting Political Priorities

In its *Environmental Report 2000*, the Federal Government's Council of Experts for Environmental Issues points out that in comparison to the early 1990s, 'the environmental issue has notably reduced significance' for the German public (*Sachverständigenrat 2000:* Section 8). A Government-commissioned survey on *Environmental Consciousness in Germany* (Preisendörfer 1998) clearly confirms this view. Throughout the 1990s, the German public has always regarded environmental issues as less urgent than unemployment, competitiveness in the world market, asylum seekers, and various other social issues (Preisendörfer 1998, Gibowski 1999, also see Baukloh and Roose in this volume). Table 7.5 shows the change in the political priorities of the electorate between 1996 and 1998.

Table 7.5 Political priorities on an *urgency scale* from 1–10 (mean values)

	1996	1998
unemployment	9.3	9.6
crime	8.6	8.7
economic competitiveness	8.3	7.9
environmental protection	8.1	7.8
welfare state	7.9	7.8

(source: Preisendörfer 1998: Section 3.1)

As Preisendörfer points out, the percentage of those regarding environmental protection as important or very important went down from 72 percent in 1996 to 62 percent in 1998. Whilst in the Western *Länder*, the environment ranges on position four, it is only in position five in the East (Preisendörfer 1998: Section 3.1). The percentage of those 'feeling concerned when thinking about the environmental conditions our children and grandchildren will probably have to live in' went down from 74 percent in 1996 to 65 percent only two years later. The figure of those believing that 'if things continue as they have been so far, we are heading straight towards an ecological disaster' went down by a full ten percentage points (1996: 66 percent; 1998: 56 percent). Furthermore, in 1998 only 50 percent believed that 'industrial countries have already exceeded or will soon reach the limits to growth' (56 percent in 1996). Bearing in mind that the limits-to-growth argument once provided the starting point for the eco-movement, this observation is particularly interesting to note. Another remarkable result of Preisendörfer's survey is that only 47 percent rejected the suggestion that 'environmentalists often strongly exaggerate the significance of environmental problems' (54 percent in 1996) (ibid.: Section 3.2). Of course, these figures also show that the level of environmental awareness and concern is still high in Germany, but it is evident that 'in the public perception, not only the relative importance of environmental protection has declined, but the level of ecological sensitivity has as well' (ibid.).

Coming back to an issue that was discussed in the previous section, the shifting political priorities of the German public can also be seen from changes in the membership of environmental organisations. Despite the encouraging absolute figures collated in Table 7.1, various studies have revealed that the percentage of people who are members of an environmental NGO has recently declined. Table 7.6 shows that young people, in particular, no longer feel as attracted by these organisations as they used to.

Table 7.6 Membership in environmental groups or organisations (in percent)

	1996	1998
age 18–30	7.7	3.6
age 31–45	5.9	5.8
age 46–60	4.9	4.0
over 60	3.4	2.9
total	5.4	4.2

(source: Preisendörfer 1998: Section 8.2)

Referring to a slightly wider spectrum of NGOs, Fischer et al. (2000: 276) found that amongst young people (aged 15–24) membership in environmental and human rights organisations declined from 3.2 percent in 1996 to 2.8 percent in 1999. Looking at various forms of individual activism for the environment, Preisendörfer comes to the conclusion that 'altogether the level of activity has declined in comparison to 1996' (1998: Section 8.2). These findings reconfirm what was said above concerning the declining appeal of green parties to young people. They are also in line with the observation that young people are generally increasingly uninterested in political matters (see Figure 7.1). The hypothesis that young people are not generally apolitical but merely losing interest in *traditional-style* politics and political institutions (e.g., Beck 1997, 2000) might need revisiting.

Figure 7.1 Young people (age 15–24) interested in politics (in percent)

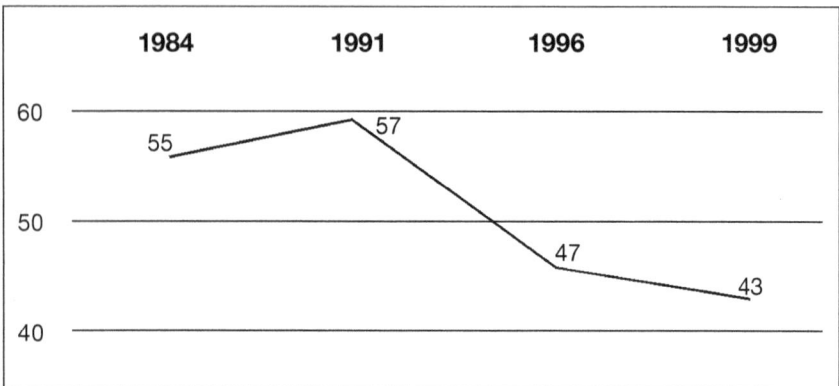

(source: Fischer et al. 2000: 263)

As further evidence for the shifting political priorities of the German public we might finally turn to public consent to the Government's legislative efforts. Table 7.7 shows that although there is still a large majority who regard environmental legislation and the actual enforcement of environmental laws as insufficient, there is a remarkable increase over the 1990s in the percentage of those regarding both aspects as acceptable. This development is particularly interesting because criticising overly lenient environmental regulations and the so-called implementation deficit has undoubtedly become a matter of *ecological correctness*. It is an easy way of displaying environmental awareness without committing oneself to any immediate consequences. The figures collated in Table 7.7 therefore probably only reflect those feeling very strongly about this issue, i.e., those very strongly against stricter legislation, whilst in real terms the level of public consent is probably significantly higher.

Table 7.7 Public support for environmental legislation (in percent)

	1991	1992	1993	1994	1996	1998
environmental legislation is satisfactory	23	27	31	32	32	37
implementation is well monitored	9	12	11	14	15	18

(source: Preisendörfer 1998: Section 9.3)

Of course, these figures give no more than an impressionistic view of environmental attitudes and public priorities. Nevertheless, they provide evidence that there is not much reason to believe that the electorate would, in the current cultural, economic and political climate, support any attempts to enforce new pro-environmental laws and lifestyle changes. The debate on fuel prices over a period of several months in 2000 and the vociferous demand to take back the ecological tax reform were a clear expression of the public mood. The suggestion that the environmental movement is going through a phase of 'strategic reorientation ... on questions of *sustainable life styles*' (Brand 1999b: 56) seems somewhat optimistic.

The Changing Function of Social Movements

This diagnosis that the general climate and political constellations are unfavourable for a revitalisation of eco-politics, and that the recent radicalisation of small segments within the eco-movement should not be regarded as indicators of a new wave of ecological mass mobilisation, can be further developed and placed on a broader sociological foundation by taking into account that in late modern societies the function of social movements has radically changed.

In the early 1970s the then *new* social movements had appeared on the political stage with the slogan and diagnosis that 'the system is bankrupt' (Kelly 1984). They saw themselves as the agents of the 'departure into a different society' (Brand et al. 1986). Particularly in Germany, they had emerged from a situation of political stand-still (grand coalition), and with their new topics and strategies they promised to bring new momentum and innovation into social and political affairs. Three decades later, the emancipatory and partic-ipatory revolution they triggered has led to an unexpected result: political protest and direct intervention have become a normality, a political means employed by all kinds of social groups. Farmers, mothers, doctors, ravers, miners, pensioners, gays, etc., all take to the streets and demand to be heard. Yet in the 'movement-society' or 'protest-society' (Neidhardt and Rucht 1993, Rucht 1999a) such well organised and effectively articulated minority interests confront politicians with tough choices and render efficient policy making extremely difficult. The permanent threat of popularity polls and upcoming elections further aggravates the problem. Thus a new form of *democratic sclerosis* or political deadlock (Beck 1993, Raschke 1999) has emerged, which for the past decade has effec-tively paralysed German politics and blocked any ambitious reform policy.[12] Whilst the minority groups are fully preoccupied with their own particular egoisms, large-scale developments and the perspec-tive for society as a whole seem to be determined by factors and driven by forces which are beyond political control.

In the context of globalisation, the drive for innovation and social change comes no longer from the bottom up, but it seems to be coming from the top down. Ever faster rounds of technological inno-vation originate from the science laboratories and electronics companies. Terrified by the idea that they might fall behind in the global race for competitiveness and foreign investment, governments are trying to respond with appropriate policies of societal moderni-sation. Societal innovation and change is thus no longer a demand carried forward by progressive protest movements, but in late modern societies they impose themselves onto society, partly in the form of systemic necessities, and partly enforced by governments desperately trying to prove that they are still in the political driving seat. In this situation protest movements no longer 'proactively inter-vene in determining change', but first and foremost they 'normatively demand a restriction of change' (Raschke 1999: 78). Rather than developing and promoting ideas for a radically different society, they fight undesirable side effects of large-scale developments, which they can hardly conceptualise, let alone control or even genuinely influ-ence. In the maelstrom of innovation that characterises late modern societies, social movements are focusing on the task of slowing down uncontrolled change and hastened policy responses, thus trying to restrict the damage that might be done by, for example, the neo-liberalist free market agenda.

Similar to some movements around the turn of the twentieth century (see Rohkrämer in this volume), societal protests and social movements thus become, once again, a retarding, often regressive, rather than progressive force. Working to the logic of the lesser evil, they prefer the respective status quo, which – although perhaps lacking in many respects – is known and calculable, and which therefore generates a sense of security. Given the reactive, security-oriented and minority-interest-centred make-up of contemporary political protest and mobilisation; and given, beyond this, the general 'crisis and even disappearance of macro-political collective actors capable of strategically coordinated action' (Raschke 1999: 74), the revitalisation of the environmental movement as a power-ful political actor seems rather unlikely. The current state of environmental politics and the eco-movement cannot simply be described as one of the periodic lows as they supposedly occur in issue-attention cycles and the life-cycle of social movements.

The Break-Up of the Ecologist Constellation

In conclusion, the ideological crisis of the eco-movement and its organisations needs further analysis. The value preferences and political priorities outlined above might theoretically at some stage shift back towards ecologist values. There is also the age-old argu-ment that contemporary society just needs 'a deeper sense of crisis' before the eco-movement and environmental politics can gain real strength and momentum (Rucht and Roose 1999: 76). In order to disprove such objections and reconfirm the hypothesis that in contemporary society ecologist values are not just being postponed but once and for all abandoned, it is useful to explore how key parameters which once provided the basis for ecologist thinking have since changed in a quite fundamental way.

The eco-ideological package has always consisted of two analyti-cally distinct dimensions which were, however, in practice inseparably intertwined: on the one hand ecologism is concerned with the physical environment and the material foundations of human and other life, on the other hand it responds to specific cultural concerns and identity needs. In ecological politics and the eco-theoretical literature the existence of these two dimensions is widely acknowledged. Yet, as ecologists have always insisted on the scientific foundations and validity of their argument, and because the cultural dimension of ecological thinking has always been difficult to capture, the quality and relative significance of cultural parameters within the ecological discourse and eco-politics remain undervalued and under-researched. Nevertheless, the rather banal truths that there tend to be no environmental politics where there is no environmental *concern*, and that *concern* may well have its origins in completely fictitious realities, make it immediately clear that green politics is not primarily about *physical conditions* or *objectively existing*[13] *problems*, but first and foremost about *human anxieties*, i.e., the violation of cultural norms and expectations.

In other words, ecological politics is to a large extent about a set of cultural values, practices and beliefs, which decide about the way in which the material world and specific physical conditions are perceived (mediated), and environmental problems (anxieties) framed, publicised and remedied.[14] For the European political ecology movement and the ideology of ecologism the tradition of Enlightenment humanism provided this normative foundation.[15] The supposedly universalist values of this tradition (rationalism, egalitarianism, liberalism, human rights) determined the specifically ecologist way of constructing nature, naturalness and self-identity. They were assumed to promote not just the collective self-construction, self-determination, self-realisation, self-control and self-experience of the rational and autonomous individual, but at the same time to secure the integrity of nature and thus the systematic coherence and equilibrium of the universe. Ecologist politics therefore implies not merely – and arguably not even primarily – the task of protecting the extra-social physical environment, but also the task of constructing a specific kind of individual identity and societal order. The so-called *ecological* problem is the inseparable unity of an *environmental* problem and a *societal* problem. Ecologist politics is the inseparable amalgamation of a *politics of material conditions* and a *politics of self-identity* (Blühdorn 2000a).

However, green issues, which started their career at the grassroots level, have gradually moved up to the national and international levels. Ecologists themselves actively promoted this development. Because grass-roots activism was often inefficient and certainly insufficient, they established their own national and international organisations. Increasingly, environmental problems have been seen as international or even global problems, which therefore had to be dealt with by international and global structures and actors. However, the gradual shift to the international level has had a considerable impact on the substance of ecologist politics. In order to facilitate international communication and consensus, ecological issues have had to be expressed in supposedly value-free scientific terms and negotiated in accordance with the universally shared economic code. In other words, ecological problems have had to be reformulated as scientific and economic problems. This has implied that the ecologist package-deal, i.e., the specifically ecologist construction of nature, naturalness and self-identity (physical conditions seen through the glasses of a particular value system) had to be untied. The internationalisation of environmental politics has meant the reduction of the 'ecological package-deal' to the scientifically and economically communicable issues now being negotiated in the globally dominant discourse of 'ecological modernisation' (Blühdorn 2000b, 2001).

At the sub-national level, this process of untying the eco-ideological package has been reinforced by the differentiation, pluralisation and individualisation of cultural values, practices and preferences. Due to ongoing processes of modernisation, culture-specific patterns of constructing nature, naturalness and self-identity are becoming ever more diversified and exclusive. They are becoming restricted to ever smaller cultural communities, eventually perhaps culminating in completely idiosyncratic preferences, interests and expectations. Accordingly, the self-determined articulation, realisation and experience of sub-culturally (personally) specified constructions of nature, naturalness and identity can no longer be coordinated through large-scale movements and organisations. It reverts to the level of more diversified and exclusive sub-cultures. In a strongly simplifying and schematic way, this two-way process of unpacking the ecological package-deal is illustrated in Figure 7.2.

Figure 7.2 The unpacking of the eco-ideological package

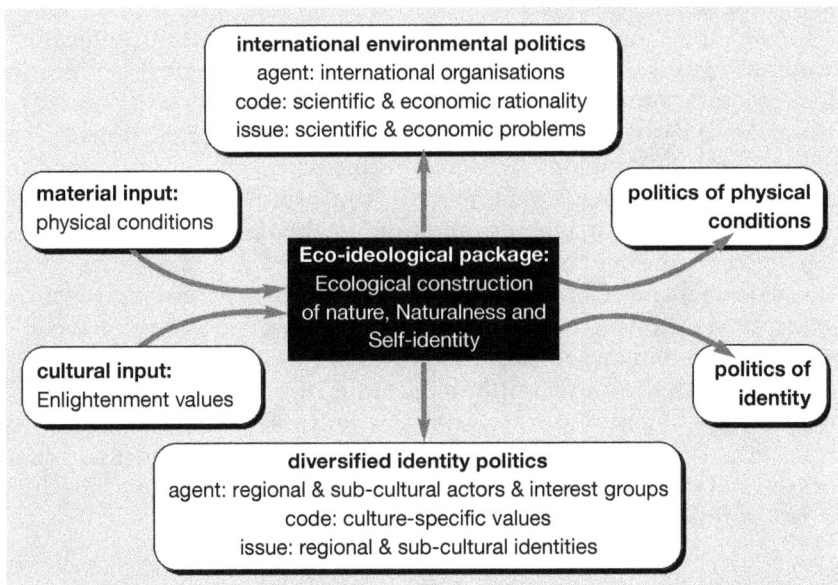

international environmental politics
agent: international organisations
code: scientific & economic rationality
issue: scientific & economic problems

material input:
physical conditions

politics of physical
conditions

Eco-ideological package:
Ecological construction
of nature, Naturalness and
Self-identity

cultural input:
Enlightenment values

politics of
identity

diversified identity politics
agent: regional & sub-cultural actors & interest groups
code: culture-specific values
issue: regional & sub-cultural identities

The pluralisation of cultural constructions of nature, naturalness and self-identity has led to an internal differentiation and fragmentation of the eco-movement. If it ever really could, the ecological issue can clearly no longer function as a generally shared basis for collective mobilisation towards an all-inclusive eco-humanist utopia. Although in abstract terms the environment remains a common concern and interest, it loses its status as a potential source of a new social consensus. The ecologist vision of a participatory, inclusive and integrative eco-politics gives way to a politics of conflicting interest groups (conflicting representations of nature and naturalness). The diversity of eco-political demands, as well as the need to reconcile eco-political with other political interests, obviously increases the pressure to reframe environmental issues as material and economic issues. In this way the reductionism of international environmental politics is reproduced at the national and sub-national levels. Phrased positively, this may be described as the integration of environmental issues into the existing political and economic structures and processes. Phrased negatively, one might say that the eco-movement has uniquely failed to formulate any genuine alternatives to the core values of capitalist consumer society (Blühdorn 2000b, 2001). For this reason, the attempt to stabilise green parties by refocusing them on their original values is doomed to failure. The suggestion that the eco-movement might avoid 'losing its soul' by bringing its values 'to the forefront of its work' rather than allowing them 'to languish in obscurity' (Porritt 1997: 71), relies on the existence of values which have long since dissolved.

Given the cultural and political parameters that determine contemporary societies, there is no space for ecological politics in the ecologist sense. The eco-ideological package has become inappropriate and so have the ecologist actors and organisations. 'The special conditions which created the original core support for the German Greens ... show little sign of returning' (Jahn 1997: 181). As was indicated in the introduction to this chapter, this implies that even in the 'century of the environment' ecologist ideas will never be realised. Contemporary societies are not moving towards 'green futures' in the ecologist sense, and there is no future for ecologist parties and organisations. However, contrary to Bramwell's prophecy this does not imply a 'decline of environmental politics in the West'. Contemporary societies will, obviously, continue to deal with their redefined environmental problems – but this redefinition entails significant changes in the substance, style and ethos of environmental politics.

Notes

1. Unless indicated otherwise, the terms *green* and *greens* are used in this chapter in a comprehensive sense, i.e., not restricted to specific party organisations.
2. The term *ecologist* here signifies a comprehensive world view and political ideology. Ecologists demand fundamental change in the most fundamental principles of contemporary society and believe that neither traditional *conservationist* approaches nor the incremental reforms favoured by *environmentalists* are sufficient to solve the *ecological crisis.*
3. The term refers to the totality of societal conditions determining the success or failure of environmental politics (e.g., level of information, environmental awareness, environmental value orientations, efficient eco-political actors, appropriate legislation and rules of their enforcement, etc.).
4. In Spring 1998 the Greens had re-tabled a suggestion originally made by the *Sachverständigenrat für Umweltfragen* (Council of Experts for Environmental Questions) in 1994 according to which the price for petrol was to be gradually increased (roughly trebled from 1998 levels) to five DM per litre. Strategically, this suggestion turned into a fiasco. Public support for this environmental measure was at rock-bottom level. For several weeks the media kept the debate going and ensured that during the run-up to the elections *Die Grünen* were primarily associated with higher petrol prices.
5. Phase 1 of the ecological tax reform came into force in April 1999. Petrol prices went up by 6 Pfennigs per litre, fuel oil by 4 Pfennigs, gas by 0.32 Pfennigs per kwh, and electricity charges increased by 2 Pfennigs per kilowatt-hour. Phases 2 and 3 (implications similar to phase 1) were implemented in January 2000 and 2001. Further increases are planned for the next three years, however, the Conservative opposition parties have orchestrated significant populist resistance against the project.
6. Following extremely long-winded negotiations with the nuclear industry, the Schröder-Government secured an agreement (June 2000) wherein the existing nuclear power plants in Germany should normally be decommissioned after 32 years in operation. Operation times, however, are transferable, and the Conservative opposition parties immediately announced they would annul this deal after any future change in government.
7. Here, and for the other organisations, in mio. DM; figures include membership fees, donations, eco-sponsoring, bequests, fines etc., as appropriate for the respective organisation.
8. Joschka Fischer: Foreign Ministry; Jürgen Trittin: Environment, Nature Protection and Reactor Safety; Andrea Fischer: Health.
9. This was justified particularly by Schröder's already lengthy negotiations with the nuclear industry (Energiekonsensgespräche) and Oskar Lafontaine's long-standing commitment to an ecologically more sensible system of taxation.
10. In June 1999 Schröder presented the so-called *Schröder-Blair-Papier* on the renewal of European social democracy, which attracted hardly any attention in Britain but caused major confusion amongst the German Social Democrats, forcing Schröder to pull back and make concessions to the Social Democratic left.
11. *Castor* are special high security containers used for the transportation of nuclear waste.
12. The long overdue reform to the German tax system as well as the reform of the health and pension systems are two good cases in point.
13. In the sense of existing independent of and prior to human experience and social discourse.
14. In order to illustrate this at first sight perhaps irritating idea, one might refer to the different ways in which different cultures think, for example, about the practice of whaling or the barren hills of the English Lake and Peak Districts. Furthermore, one might ask why the killing of whales, seals and dolphins triggers significant waves of protest, whilst the disastrous depletion of cod, haddock or anchovy populations raises hardly any public concern.
15. To the extent that the so-called deep ecologist currents display anti-humanist elements, they remain exempt from this generalisation.

PART IV

Literary and Filmic Discourses on
the Environment

CHAPTER 8

The Great Blind Spot[1]

Carl Amery

Must we become inhuman in order to save humankind?[2] Hitler answered the question, both in his political programme and as a practical politician, with a resounding yes. Not only did he face up to the demands of life, as he saw them – he also invested life with mystical meaning, as the cruel queen,[3] whose marshal and execu- tor he undertook to become, in order to preserve the human species. (He referred to this explicitly as his supreme political aim.) In declaring the Jews the arch-enemies of sustainability, he took on the whole Jewish-humanist message – a message of peacefulness, of the preservation of weak and damaged forms of life, of the neces- sity for discussion and compromise. He was able to win the support of a disillusioned people because his ideas were rooted in the think- ing of his time, and because he promised the German people that, by applying his formula, they would themselves become the queenly rulers of the human race and the leaders of an advanced human civilisation – with a right to all the associated privileges.

He failed, because he understood little of the actual inner state of the world, of the democratic western world in particular, because he strove in hectic hypochondria to squeeze plans spanning the century, indeed the coming thousand years, into the space of his own lifetime, and because Germany was too slender a power basis to support such plans. Yet he believed he could still attain his main aim, carry out his central task in fealty to the cruel queen, and make his contribution to natural history: annihilating Judaism was an essential precondition for the extinction of the humanist message, for the abolition of the last and greatest of threats to the world.

Must we become inhuman in order to save humankind? Pol Pot was certainly in agreement with Hitler. Stalin and Mao Zedong too in essentials. Without openly admitting it, their political actions led in the same direction. These inhuman dictators sought an answer to a question which the world of the Atlantic Charter, the world of the great western empire, in which we are still at home, has studiously avoided: Can humanity survive its own achievements? And what is the price of such survival, in terms of comfort, dignity, human rights and self-determination?

Hitler Redivivus?

The plans of all the dictators were, mercifully, flawed by oversights and contradictions, and humanity has shaken them off. But until we find a more human answer to Hans Jonas's burning question, Dracula lives on down in the vaults, slumbering under the rubble.

In fact he has already been reappearing in new ghoulish disguises in the last years of the millennium: as the barbaric chief of a would-be master race, as fundamentalist killer, and as ice-cool planetary manager with superior, discreeter methods of surveillance and selection. In thousands of roundabout and unobtrusive ways these methods are already being put into practice today, and the explicit or tacit acquiescence of the public is always a trade-off, a *commercium*: liberty and dignity are the price paid for security. Once the hollowness of this security is recognised in view of the ecological threats to the future of human society, the fragility of the human rights founded on the UN Declaration, and of our postulated right to freedom from fear and need will become apparent.

Only then will the enormity of Auschwitz as a historical phenomenon become evident. For it was no natural catastrophe outside the orderly progress of history, but the primitive anticipation of an option open to realisation in the century now beginning. The 'rational discussion and analysis of the past' we are urged to engage in will then reveal itself as wholly inadequate. Behind the cobweb-festooned files, the boxes of photos, the outdated commentaries and concepts, the question of Hitler's real meaning will appear in a new light. And the well-meaning assurances so often heard in the debate on how it all came about, that Germany today has come such a long way since those terrible years, that it has become democratic, tolerant and self-critical, will lose their relevance. For it will no longer be a question of a mere local application of the Hitler formula in central Europe. It will rather be a matter of its contemporary relevance in the twenty-first century – and our ability to avoid it.

What will become of the humanist message?

The planetary manager of the future will be at one great disadvantage in comparison with Hitler: it will be more difficult for him to identify the arch-enemy, the deadly virus,[4] and to attack a scapegoat, a ritual victim. The hated virus was, as we have seen, ultimately a message; the message of the possibility of overthrowing the cruel queen, of protecting the weak and the disadvantaged from her aristocratic principle, of emancipating human actions from the brutish laws of nature.

This message was elevated to the official doctrine of humanity in the decades of the great moratorium[5] – not least because the atrocities committed by the Nazis were still fresh in people's minds. But we will not be able to avoid the question of sustainability in the long run. What will be the outcome of the clash between the humanitarian message and the laws of nature, when the credit we are living on in terms of energy and natural resources is used up, and the artificial world of modern civilisation comes to an end? Will we then find ourselves face to face with the multi-facetted eyes of the cruel queen – or will humankind as a whole succeed in taking nature and its laws as a *cultural* challenge to us all, as an invitation to develop a new understanding of life, to reach maturity for the first time, and enter into a symbiosis with all the fundamental facts of life? Without delegating suffering and death to a 'Master from Germany'?[6]

Hitler's solution was a ghastly cheap trick, the illusion of modernised barbarism: life as a master race at the expense of everyone else, as a privilege in return for the preservation of the species, couched in terms of a thousand-year Reich. His rule was short, it was unable to prove itself, but the validity of his formula was not conclusively disproved either. There is a danger that, purged of the absurdities of the figures he entered into the equation, and aided by a more advanced knowledge of the mechanisms and techniques of control and domination, this card may be played again for all it's worth in times of much more serious crisis – and that Hitler's old formula will initially only be recognised by the most astute.

We need a completely new concept of culture to resist it effectively, one based on a fundamentally different formula. For it is ultimately a matter of how we understand the world of living things in which we exist as 'life which wants to live' (Albert Schweitzer).[7] Death is a fundamental fact of this life. To ignore this is foolish and dangerous. But the alternative is not a battlefield in which everyone is pitted against everyone else. No species survives through the death of all the others, nor can a master race; that was and is the weakness of Hitler's vulgar Darwinism. Nor can we continue to observe the old naïve Jeffersonian formula of innocent usufruct.[8]

What we need to develop is a new solidarity with the biosphere, the world of living things, purged by knowledge and humility. The Darwinism of the neo-cannibals, regardless of whatever ilk, has no more place in it than the naïve doctrine of the invisible hand of providence or the arrogant hope in eschatological redemption by a force above and beyond us still discernible in many a secular mind.

If there is any point in seeking a global formula today, then it must be the following: *Humankind can remain the crown of creation – as long as we understand we are not that by right.*

Notes

1. The final chapter of *Hitler als Vorläufer. Auschwitz – der Beginn des 21. Jahrhunderts?* (Hitler as a Precursor. Auschwitz – the Beginning of the 21st Century? Amery 1998: 187-191), is here translated with kind permission of the author.

2. The question is taken from the eco-philosopher Hans Jonas, who expresses repeated doubts about the ability of democracy to cope with the ecological challenge in *Das Prinzip Verant-wortung* (The Principle of Responsibility: Jonas 1979: 55f., 262-5, 3 40). Jonas argues that forms of tyranny are at a natural advantage in imposing a regime of asceticism and curbing population growth.

3. See Chapter 4 of Hitler's *Mein Kampf* (Hitler 1935: 144).

4. The term is used by Hitler to describe the Jews in *Mein Kampf* (Hitler 1935: 62).

5. In an earlier chapter Amery refers to the years of freedom from fear and want in Europe following the Second World War (approximately 1950–1970) as a 'moratorium', when the threat of the exhaustion of natural resources was not yet apparent. Europeans were, as he points out, in fact living at the expense of other continents and future generations.

6. The phrase is taken from Paul Celan's famous poem 'Todesfuge' (Death Fugue). See Paul Celan *Gesammelte Werke in Fünf Bänden*, ed. Beda Allemann and Stefan Reichert. Frankfurt am Main, Suhrkamp, 1983, vol. 1, 41f.

7. Albert Schweitzer (1875–1965), missionary doctor, protestant theologian, philosopher, organist and musicologist, who was awarded the Nobel peace prize in 1953, propounded an ethic based on reverence for all forms of life, reflecting a passionate concern for the welfare of animals. Amery interprets Schweitzer's shift away from anthropocentrism in ecological terms. The phrase 'life which wants to live' is taken from *Kultur und Ethik* (Culture and Ethics, first edition 1923), in *Gesammelte Werke in fünf Bänden*. Vol. 2, Munich, Beck, 1974, 377.

8. Thomas Jefferson, draftsman of the American Declaration of Independence and third President of the United States, idealised the independent farmer as the backbone of a healthy republic, thus formulating a central aspect of the national self-image. In Chapter 8 of *Hitler als Vorläufer* Amery mentions working in the South Reading Room of the Adams building of the Library of Congress (built in 1938), which is decorated with murals based on themes drawn from quotations from Thomas Jefferson's writings. The south wall bears the following inscription: 'The earth belongs always to the living generation. They manage it then, and what proceeds from it, as they please during their usufruct. They are masters too of their own persons, and consequently may govern them as they please. (Jefferson to James Madison, September 6, 1789)'

CHAPTER 9

Writing Environmental Crisis: The Example of Carl Amery

Axel Goodbody

'Writers are always, by definition, preservers of nature' (Schiller 1962: 432)

Political Activism and Literary Commitment

Amery's standing in Germany as a critic, essayist and novelist is reflected in a string of literary prizes. He has served terms as elected president of the German Writers' Union and the German PEN, and recently been the subject of a major exhibition and accompanying publication in Munich. A founding member of the Green Party and author of influential commentaries on environmental issues, he has meanwhile enjoyed a national reputation as a leading Green thinker for the last quarter of a century.[1] Yet since his early study *Capitulation: An Analysis of Contemporary Catholicism* (Amery 1967) none of Amery's books have been translated into English, and he remains one of the least known in Britain of the generation of politically committed writers who dominated West German literature in the second half of the twentieth century, of whom Heinrich Böll, Günter Grass, Peter Weiß, Hans Magnus Enzensberger and Erich Fried were among the more prominent.

Carl Amery, whose real name is Christian Mayer – the anagrammatic pseudonym 'Amery', adopted to avoid confusion with other literary Mayers, has, ironically, led to occasional confusion with the Austrian Jean Améry – is a German intellectual. That is to say he is one of the elite of authors, historians, philosophers, sociologists, psychologists and literary critics with a high profile who have spoken out on issues of social and political importance from non-conformist, usually liberal-left positions outside the power structures of the country, contributing to intellectual debate in the media and constituting a 'conscience of the nation'. The term he has used to describe himself is 'moralist'. Placing himself in the tradition of the seventeenth and eighteenth-century French *moralistes* de La Bruyère, Montesquieu and Voltaire, radical thinkers who coupled logical argument with the barbs of irony and the subversive power of fiction in searching critiques of the political and social structures of their time, he has also cited Gotthold Ephraim Lessing, the German Enlightenment humanist, as a role model (Kiermeier-Debre 1996: 184). The political, social and cultural concerns of such authors, however, require fundamental reformulation in the light of the new biospherical responsibility imposed on us by modern technology, he writes in his essay 'Macht und Ohnmacht der Literatur im Zeitalter der biosphärischen Verantwortung' ('The Potential and Impotence of Literature in the Age of Biospherical Responsibility' – Amery 1991: 265–271). The foremost task of literature in our age has become one of 'interpreting and clarifying the ultimate, supreme [ecological] responsibility of man in social, political and cultural terms' (269).

Besides environmental commitment, politically engaged Christianity and (critical) Bavarian patriotism have been constants in his writing. Born in Munich in 1922, he grew up in a milieu of enlightened Catholicism and was educated at a traditional grammar school (*humanistisches Gymnasium*). His father, Professor of History and the History of Art at the *Philosophisch-Theologische Hochschule* in Freising (near Munich), was an expert on the liturgy and involved in liturgical reform. He was a personal friend of Carl Muth, editor of the journal *Hochland* (Highland), and Amery came into contact with the contributors to this cultural forum of conservative opposition to National Socialism from an early age. His first literary efforts included press reports and poems in local newspapers. At the age of sixteen he compiled a typed collection of detective stories in historical and contemporary settings (ancient Egypt, Rome, Capua, Paris and North America) under the title *Das Rad* (The Wheel). The manuscript (extracts of which are published in Kiermeier-Debre 1996: 20–29) contained an introduction distinguishing between different types of detective story, in which the author acknowledges the influence of G.K. Chesterton's *Father Brown stories*.[2] The booklet already reveals Amery's characteristic blend of learning and popular appeal, entertainment and morality.

Called up in 1941, Amery served in the *Wehrmacht* in France, Russia and North Africa. His capture by American forces in Tunis in 1943 was an event of crucial importance for the rest of his life. He spent the next two years as a Prisoner of War in Texas and Arkansas. On release in 1946, he returned to Munich, enrolled as a student and began to publish stories in the new realistic style about his experiences during the war. However, he was able to revisit America for a year 1948/9 with a scholarship to study literary theory and criticism at the Catholic University of Washington, DC. The links he forged in the United States were of lasting personal, political and literary importance: he married an American in 1950, wrote for several years as a correspondent for American newspapers in Munich, and took a keen interest in intellectual and cultural developments in the United States. American influence is unmistakable in both the style and content of the satirical novels he wrote in the 1950s, which sought to convey a serious critique of contemporary society in a form accessible to a wide readership, and in his later best-selling novel *Der Untergang der Stadt Passau* (The Fall of the City of Passau – Amery 1975) borrows from American science fiction. In addition, his familiarity with the environmental movement in America, which preceded the German movement by a decade, including the alarmist debate sparked off by Paul Ehrlich and Barry Commoner in the late 1960s, explains how he came to be at the forefront of developments in Germany in the 1970s. His speeches and essays on environmental issues frequently allude to American history, and contain references to and quotations from Rachel Carson, Aldo Leopold and other US environmental writers, while the novels *An den Feuern der Leyermark* (The Camp Fires of the Leyermark – Amery 1979) and *Das Geheimnis der Krypta* (The Secret of the Crypt – Amery 1990) build fictional bridges between German and American history and politics. American Indians, who had of course captured the imagination of previous generations of Germans discontented with European civilisation, from the Romantic Johann Gottfried Seume to Friedrich Engels and the popular adventure story writer Karl May, are a constant point of reference in Amery's novels and essayistic reflections on a way of life in tune with the earth and the biological foundations of human society. As an individual Amery thus reflects the convergence of native and foreign elements present in the German environmental movement, whose roots lay in the German tradition of civilisation critique, but whose dramatic rise in the 1970s owed much to the American example.

From 1950 on Amery made his living as a cultural journalist, publishing political commentaries in the press, reviewing books and films, and assessing film projects and American and British novels for translation. In the second half of the 1950s he worked increasingly for radio, penning historical portraits, features and plays, and writing scripts for TV drama. His first two novels, *Der Wettbewerb* (The Competition – Amery 1954) and *Die Große Deutsche Tour* (The Great German Tour – Amery 1958), presented a trenchant but polished and humorous critique of the German economic miracle. They gained him recognition and admission to the influential writers' association *Gruppe 47*, but also the reputation of a 'satirical talent', which was to stick long after it ceased to do justice to his writing. The critical study *Die Kapitulation, oder deutscher Katholizismus heute* (Capitulation, or German Catholicism Today – Amery 1963) placed him at the centre of public controversy. His analysis of the Catholic milieu and the role of the church in the Third Reich was denounced from the pulpits but praised in *Der Spiegel*, and sold 100,000 copies. Amery was asociated in the public mind with Heinrich Böll, the critical social chronicler of postwar Germany in novels such as *Ansichten eines Clowns* (The Clown, 1963), and Rolf Hochhuth, whose play, *Der Stellvertreter* (The Representative, 1963), accused the Pope of indifference to the holocaust. It led to a longstanding friendship with the Nobel prize-winner Böll, with whom Amery was united in opposition to the postwar restoration from a standpoint of left-wing-liberal Catholicism. 'Plus und Minus, niemals Null' ('Plus and Minus, never Zero'), an autobiographical manuscript dating from the mid 1970s meditating on humankind's self-induced separation from nature, traces the continuity of Amery's moral and political commitment from antifascism through the critique of postwar Catholicism to environmentalism (Kiermeier-Debre 1996: 129–144).

The second half of the 1960s was a period of predominantly extra-literary activity for Amery. He was appointed director of the Munich municipal libraries in 1967, and spent three years in charge of a staff of four hundred. Having become a party member of the SPD in 1966, in the election campaigns of 1969 and 1972 he was one of the many prominent public figures, including writers (Grass, Lenz, Böll), TV personalities, artists (Grieshaber, Staeck), film directors (Schlöndorff, von Trotta), journalists, academics, actors and even sports people who volunteered their services in support of Willy Brandt in the *Sozialdemokratische Wählerinitiative* (Social Democrat Electoral Initiative), explaining and popularising the new policies and ecological initiatives developed by politicians such as Egon Bahr, Erhard Eppler, Freimut Duve and Johano Strasser. At the end of the decade he became the SPD's most outspoken campaigner on environmental issues. 'Alarm im Raumschiff' ('Alarm on the Spaceship'),[3] a speech delivered in the 1970 elections to the Bavarian parliament, anticipated many of the central themes in his later essays (Kiermeier-Debre 1996: 74–86). Leaving the SPD in 1974, when Brandt was replaced by the economic realist Helmut Schmidt in the recession following the oil crisis, and Eppler was forced out of the Ministry for Development, Amery gravitated naturally towards the Green parties which were first formed at local level in the second half of the 1970s. In 1978 he campaigned in the Bavarian elections for the Greens, and in 1979 he stood as their candidate in the elections to the European Parliament.

Meanwhile Amery had become a member of the *Gruppe Ökologie*, a loose association of leading conservationists and concerned scientists, and begun to contribute articles to Horst Stern's ambitious environmental magazine *Natur*. His reputation as an environmental thinker rests principally on two books, *Das Ende der Vorsehung* (The End of Providence – Amery 1972) and *Natur als Politik* (Nature as Politics – Amery 1976). Besides these extended environmental essays, to which I return below, he contributed to the debate on nuclear power, the most contentious issue of the decade, with a study of alternative sources of energy (Amery et al., ed. 1978). At the same time, the ideas and arguments in these books were reflected obliquely in two novels, *Der Untergang der Stadt Passau* and *An den Feuern der Leyermark*. Here and in *Das Königsprojekt* (The King Project – Amery 1974), Amery came into his own as a creative writer. He succeeded in integrating profound concerns about our relationship with nature and the future of humankind into imaginative and skilfully constructed narratives, projecting the problems of the present into a fictional future or past, and drawing on his intimate knowledge of Bavarian history and culture for colourful detail.

Der Untergang der Stadt Passau, published in paperback in the Heyne science fiction series and on sale not only at bookshops, but also at kiosks all over the country, is currently in its seventeenth printing. Its exciting plot, accessible language and humorous allusions to contemporary life have made it a success with generations of teenage readers, especially in Bavaria, where the action is set. The story, which has been adapted for the stage and performed in schools, has also been studied in sixth forms as a text rewarding examination both for the issues it raises and for its literary qualities. (See Leiner 1981 and 1987.)

The influence of Walter M. Miller jr's trilogy *A Canticle for Leibowitz* (Miller 1960), an elaborate post-catastrophe narrative inspired by the nuclear arms race, is discernible in the time scale, narrative structure and individual motifs of *Der Untergang der Stadt Passau*. However, Amery shifts the cause of the catastrophe from nuclear war to a mysterious epidemic explained variously as the punishment of God, the work of rogue scientists and an indirect consequence of overpopulation, resource depletion and the breakdown of public order. He also modifies Miller's historical pessimism in an open ending. Hungarian nomads, the village community of the Rosmer (Rosenheimer), and the citizens of Passau exemplify the three stages of the development of human civilisation discussed by Amery in the second chapter of *Natur als Politik*. His protagonists, the Rosmer Lois Retzer and his adoptive son Marte, seek to re-establish civilisation on a surer footing after the catastrophe (which is set in the year 1981). Their efforts to 'reinstate natural conditions' are far from plain sailing. 'Der Scheff', who is attempting to reconstruct the former technology-dependent civilisation in the city of Passau, must first be outwitted, and his descendants defeated, before the 'salt empire', centred on Salzburg, is eventually founded in the final pages – a social and political entity which may well eventually repeat the errors of the past.

Der Untergang der Stadt Passau does not, however, exude the apocalyptic pessimism of the spate of later German post-catastrophe novels penned by Matthias Horx, Udo Rabsch and others in the early 1980s, in the shadow of acid rain, nuclear rearmament and sabre-rattling between the superpowers. The Rosmer, led by Lois Retzer (who, we are told, was working on a theory of 'ecological materialism' reminiscent of Amery's own *Natur als Politik* before the catastrophe), constitute a (schematic) model for an ecologically stable society. They lead a relatively 'natural' way of life, as 'cooperatives' of semi-nomadic hunters, gatherers and sporadic crop-raisers, and practise grass-roots participatory democracy. Compared with classics of the green utopian genre such as the American Ernest Callenbach's *Ecotopia*, which was written in the same year (Callenbach 1975), Amery's novel does not, however, develop such ideas, and is only peripherally concerned with the form a future society based on the principle of sustainability could take.

In the 1980s Amery participated in the (successful) protest against the siting of a nuclear reprocessing plant near Wackersdorf in rural north Bavaria, and the (unsuccessful) campaign against the building of a new airport for Munich at Erding. These activities feature, alongside broader environmental concerns, in two major novels which appeared in this decade, *Die Wallfahrer* (The Pilgrims – Amery 1986) and *Das Geheimnis der Krypta*.

Die Wallfahrer is Amery's most ambitious novel. Readers willing to meet the challenge of its unusual subject matter (a critique of the political subversion of Bavarian piety) and linguistic demands (lengthy passages are written in a pastiche of seventeenth and eighteenth-century German) are rewarded by fascinating historical insights and a good deal of humour, alongside serious reflections on the purpose of life and the future of humanity. The structure of the narrative may be described as postmodern, in that a network of motifs and symbols constitute a constantly shifting play of allusions, perspectives and revelations. The four principal narrative strands, set in different centuries, are united by a geographical focus on Tuntenhausen (a minor Bavarian centre of pilgrimage), the protagonists' shared vision of moral corruption and divine punishment, and their attempts to put their perceived mission of religious, social and political regeneration into practice.

Die Wallfahrer may be compared with Günter Grass's major environmental novel *Die Rättin* (The Rat), published in the same year, in that both books present apocalyptic scenarios of the end of world, seeking to shock and warn their readers, leaving open the question of mankind's ability to make the shift of consciousness needed to avert catastrophe. Grass's novel, which also combines diverse narrative strands in a complex structure, appears, despite its brilliant satirical passages, at times crudely didactic in its explicit treatment of forest dieback, marine pollution and the dangers of nuclear warfare and genetic engineering. *Die Wallfahrer* focuses on morality rather than specific environmental issues. Amery revives pilgrimage as a central metaphor for our existence on earth, linking Baroque religious fervour, eighteenth-century counter-Enlightenment religiosity, pious nineteenth-century attempts to found a progressive popular rural Catholic party and misguided twentieth-century political initiatives. A chapter entitled 'Paths of Modernity' tackles euthanasia in the Third Reich and takes us up to the postwar decades in a blistering satire on contemporary Bavarian politics which caused a furore at the time.

There are two 'alternative' endings to the novel, rounding off the sequence of the historical protagonists' apocalyptic visions and underlining the author's affinity with them. In the first, a young man in jeans ascends into the heavens sounding a silver tuba – a modern angel sounding the last trump. However, Amery has no confidence in the relevance of such a conception of the Last Judgement to our need to reach the decisions required of us by the environmental situation. There is therefore a second ending, described as 'heretical', which takes place fifty million years from now. Humanity's self-destruction in the late twentieth century has not been accompanied by any Day of Judgement, and conventional religious concern with leading a good life is by implication meaningless. Gaia, the Earth Goddess, takes the place of God the Father. Amery presents Gaia as a modern reincarnation of the principle of female spirituality traditionally incorporated in the Virgin Mary, whose images and counter-images in Bavarian Catholicism, ranging from the triumphant Madonna of the imperial troops in the Thirty Years' War to the asensual ideal of the nineteenth century, have been explored throughout the book. Gaia invests transient human life with meaning by embracing both the implacably stern laws of nature and a humanist ethos of love and forgiveness:

> I believe that we must try, as I have at the end of the novel, to develop a female spirituality, to find a female figure at least half way matching the magnitude of our problems. I see this figure already to a great extent in Gaia: she is a symbol for the total complexity of life, for a biospherical complexity of life which mediates between the terrible grandeur of the laws of thermodynamics and our human existence. (Amery 1987: 16)

The American ecocritics Annette Kolodny and Patrick Murphy have analysed the problems associated with the sex-typing of nature, which can, for instance, serve to consolidate undesirable social conventions of gender-atttribution and deepen the dualist divide between the human and the natural spheres. However, Gaia, a fierce maiden rather than a benevolent mother in James Lovelock's hypothesis (where she is envisaged as a powerful, planetary-sized ecosystem capable of self-regulation and survival, if necessary by shaking off humanity), can in Amery's view serve as a useful counter-image to the resourcist concept of nature capable of mobilising readers.[4]

In Amery's last novel, *Das Geheimnis der Krypta*, ecological concerns, including distrust of the political ends to which science and technology are often put in the modern world, are conveyed comparatively directly, and this time in a popular, accessible form. Elements of autobiography, apocalyptic fantasy, ecological and political commentary are integrated in an exciting detective story. The protagonist stumbles on a global, American-led plan by a group of scientific experts to develop a biological weapon which would reduce the world population to five percent of its present size, and permit the survivors a new start. He then discovers the 'idealistic' intentions of the original proponents have been subverted by a Balkan state, which has developed an antidote to the serum and is producing stocks sufficient for its own population, in order to settle old scores with its neighbours. Like Amery's other novels, *Das Geheimnis der Krypta* clothes challenging intellectual ideas with humour and imagination: progress and scientific method are questioned in reflections on chaos theory and 'sphagistics', a 'science of the defeated' invented by Amery, whose aim is to recover the neglected potential of theories which have been historical failures.

The themes of these novels are revisited in a trilogy of science fiction radio plays and the volume *Die Starke Position oder Ganz normale MAMUS* (The Strong Position or Quite Normal MAMUS – Amery 1985b ['MAMU' standing for 'Male Academic with a University Degree']), a collection of satires cast in the form of fictional lectures, discussions and newspaper articles. Amery has published no further fiction in the 1990s, but continued to reflect on politics, philosophy and religion in the essays *Die Botschaft des Jahrtausends* (The Message of the Millennium – Amery 1994) and *Hitler als Vorläufer* (Hitler as a Precursor – Amery 1998), both of which are discussed below. Approaching his eightieth year, he continues to write for the press and speak on environmental issues at public meetings. A recent article in *Die Zeit*, drawing attention to a new publication on solar energy (Amery 1999), shows his continuing practical concern with energy policies.

Environmental Non-Fiction

Amery's non-fiction writing on the environment began in the 1970s with theological and philosophical essays exploring the root causes of our environmental problems, and shifted to politics, including suggestions for practical solutions, only to return in the 1980s to the underlying ethical questions. His essays, speeches and articles are characterised by a stern moral seriousness, but also by rhetorical eloquence and wit. Drawing on history and anthropology as often as scientific ecology in support of his arguments, they impress through the range of cultural and intellectual reference. Provocative aphoristic phrasing attracts the attention of readers, humour and irony elicit sympathy, and striking images enlist powerful cultural associations in the environmental cause.

Das Ende der Vorsehung (whose full title is The End of Providence. The Merciless Consequences of Christianity) appeared shortly after the landmark environmental publication *The Limits to Growth* in 1972, the year of the Stockholm UN Conference on The Human Environment, in which the environmental movement may be said to have begun in Europe. Amery had been following a dialogue between Marxists and Christians on the ecological bases of human life hosted by the Catholic *Paulus-Gesellschaft* at a series of conferences in Czechoslovakia. Most of his book was written before the publication of the Massachusetts Institute of Technology scientific team's study commissioned by the Club of Rome. His central argument is that Judaeo-Christian tradition has been responsible for the destruction of the environment in modern Europe in replacing the integration of human interests in natural cycles, which may have been fostered by the old nature religions, with a new attitude of human superiority. The domination of nature begun here was taken further by the Enlightenment, American Puritanism and modern experimental science. Amery's thesis, which echoes Hegel, Weber and Bloch, was not entirely original. It was preceded by Lynn White Jr's groundbreaking article 'The Historical Roots of our Ecologic Crisis' (White 1967), and Amery himself has spoken of the influence of Harvey Cox's *Secular City* (Cox 1965). The achievement of *Das Ende der Vorsehung*, which discussed many of the themes of Amery's later environmental essays for the first time, lay above all in its bold questions, stimulating comparisons, command of historical detail and radical conclusions.

Amery uncovers striking similarities between the assumptions regarding humankind's relationship with nature underlying Christianity and Marxism, in particular their separation of man from nature, their encouragement to dominate and exploit in the interest of the human species, their conception of history as leading from original paradise to a future state of harmony, their promise of ultimate stability of the biosphere (e.g., in Genesis 8 and 9), and the shared 'principle of hope' (Bloch) which gave rise to the ideology of progress. His familiarity with church history comes into its own in passages giving insight into the tensions between different currents in Christianity, between orthodoxy and heresy, between the Benedictines' concern with organisation and efficiency on the one hand, and the scorn for the material aspects of life shown by the apocalyptic prophets of the early church on the other, between the *Realpolitik* of continuity and the unworldly asceticism of mystics like St. Francis of Assisi.

Amery outlines a position of 'true conservatism' (Amery 1985a: 111), accepting limitations on our actions in order to preserve what is of value for the future. The concept of this radical conservatism, which has less to do with the policies of contemporary Conservative parties than with Erhard Eppler's *Wertkonservatismus* (conservatism of values – see Eppler 1975), is later developed further in the article 'Das Schicksal des deutschen Konservatismus und die neuen sozialen Bewegungen' ('The Fate of German Conservatism and the New Social Movements' – Amery 1991: 30–49), Amery's Oration for Heinrich Böll (Amery 1991: 318–320) and the novel *Die Wallfahrer*.

The second part of the book, entitled 'Conclusion and Challenge', reflects on the chances of the revolutionary shift in attitudes which Amery regards as essential for the long-term survival of humankind. For Amery, Paul Ehrlich's focus on the 'Population Bomb' (Ehrlich 1968) is less significant than the doctrine of economic growth and the attitude underlying our exploitation of natural resources. His sympathies lie unmistakably with the Catholic Revival writers' consciously old-fashioned ideals of purity and humility, though he recognises these are only likely to appeal to a minority. He concludes with a stern assertion of the restrictions on individuals' actions which are likely to be dictated by the need to husband resources in the future, calling for a new way of life based on voluntary renunciation and an ethic of planetary responsibility, in which traditional morals are subordinated to the dictates of collective survival. Educational programmes must be set up with urgency, and new rewards and incentives devised, if we are to end population growth and economic growth, and live in partnership with all living things (Amery 1985a: 186–196).

The lengthy reviews[5] published in the national and regional press and in specialist journals, and broadcast on radio, were overwhelmingly positive, and *Das Ende der Vorsehung* was greeted as a stimulating synthesis of factual research and speculative thought, dealing with pressing issues intelligently and imaginatively.

The impact of Amery's book, which foreshadowed Hans Jonas's *Prinzip Verantwortung* (The Principle of Responsibility: Jonas 1979) in arguing the need for a meaningful organisation of our sublimations and renunciations around a future-orientated goal in a new environmental ethics, was undoubtedly enhanced by the liveliness of the general presentation. The literary qualities of the fictional interludes 'Moby Dick in Marienbad' and the 'Word of the Absent God' (Amery 1985a: 172–6 and 196–8) were seen to convey Amery's central points particularly effectively.

Natur als Politik (full title: Nature as Politics. The Ecological Opportunity for Humanity), published four years later, is Amery's most overtly political book. Coming at a time when the first wave of grass-roots environmental activism had reached its peak, and the need for philosophical principles underpinning broader policies on the environment was becoming apparent, it was not without influence on the political programme of the emerging Green Party. *Natur als Politik* inevitably attracted the opprobrium of political opponents – Amery was denounced by a reviewer in the conservative weekly *Die Welt* (16.9.1976) as a 'demagogue', 'short on technical knowledge, strong on arrogance, and lacking creativity'. Even sympathetic reviewers detected touches of technophobia, unsubstantiated claims, an irritating failure to acknowledge the source of his ideas, and gaps in his account of ecological countertraditions. In addition, some of the perspectives for practical action mapped out towards the end of the volume, which added to the usefulness of the book at the time, today seem oversimplified or have simply been overtaken by subsequent developments.

Natur als Politik begins by reviewing the arguments of the opponents of economic growth in contemporary Germany – the Club of Rome, Erhard Eppler, Hans Magnus Enzensberger, Herbert Marcuse, André Gorz and Wolfgang Harich. (Amery's familiarity with American and British theorists such as Gordon Rattray Taylor, Harvey Wheeler and Edward Goldsmith is also evident.) The most innovative and important passages in the book are concerned with the 'Ecological Materialism' Amery puts forward as 'a lead science contradicting anthropocentrism, integrating humankind in the network of planetary relationships' (Amery 1985a: 224). We have been acting as if we were exempt from the laws of nature, he argues. We now need to transform our cultural values so as to acknowledge our creatureliness, recognise the value of toil and strain, and participate willingly in natural cycles. Critics have suggested that the axiomatic basis of this philosophy, which appears to demand equal status for all forms of life, is unsound – Amery gives no satisfactory reason why nature should be held to possess a value or rights other than those assigned it by humans. On a practical political level, his

ecological materialism also ignores a whole range of social mechanisms and issues demanding attention. Furthermore, as Thomas
Spring has pointed out, Amery himself is no materialist. In the last
two pages of the book he abandons the philosophy of ecological
materialism summarised in his eleven theses (pp. 364-7) for the religious conception of nature as divine creation he had so eloquently
expressed in the 'Wort des Abwesenden Gottes' at the end of *Das
Ende der Vorsehung*. The pathos of his account of the suppression of
nature derives less from materialism than from theological intimation, and Amery implies that without a religious interpretation of
existence there can be no ecologically correct relationship with the
environment (Spring 1986).

Despite these shortcomings, *Natur als Politik* remains worth
reading today. As in *Das Ende der Vorsehung*, the breadth of reference is stimulating, and there are many interesting observations on
the historical roots of environmentalism. A more systematic
overview of the latter may be found in a slightly later article 'Die
philosophischen Grundlagen und Konsequenzen der Alternativbewegung' ('The Philosophical Foundations and Consequences of the
Alternative Movement': in Lüdke and Dinné, ed. 1980), which
throws light on Amery's understanding of his own links with the
past. Amery was one of fifteen prominent Green activists (including Herbert Gruhl, Holger Strohm, Petra Kelly and Baldur
Springmann) who contributed to a semi-official presentation of the
Greens' aims in the book *Die Grünen. Personen – Projekte – Programme*
(The Greens. Personalities – Projects – Programmes: Lüdke and
Dinné, ed. 1980). The volume, published within months of the
founding of the party, opens with Amery's essay (pp. 9-21). Socialism, scientific ecologism and 'civilisation criticism' are identified as
the three principal roots of the German environmental movement.
The first of these goes back to the nineteenth-century socialist and
anarchist radicalism of Fourier, Kropotkin and Bakunin, whose
influence is traced in American populism, the Paris Commune of
1871, and Latin American populist movements. The second root,
scientific ecology, relaunched the environmental movement with a
new scientific paradigm. (Another recent dimension is detected in
the scientific scepticism of Paul Feyerabend, who challenges technocracy, anthropocentrism and rationalism.)

It is the third and oldest root, however, to which Amery devotes
most space. He traces the tradition of conservative cultural criticism, the late nineteenth- and early twentieth-century response to
industrialisation examined above by Thomas Rohkrämer, right
back from the *Renouveau Catholique* (Bernanos, Chesterton) to the
Jacobite risings of 1715 and 1745, and James II's ill-fated coalition
of the old noble families with Protestant dissenters in the 1680s,

in opposition to the rising commercial and industrial middle class. The aesthetic and political radicalism of Ruskin and Morris is, according to Amery, the nineteenth-century successor of this political movement. He cites the Bavarian Romantic nature philosopher Franz Xaver von Baader as a German proto-Green, and notes that, as in England, the most significant resistance to the forces of environmental destruction in the nineteenth century may be found in the aesthetic sphere, i.e., in Romanticism and the Youth Movement. (Jonathan Bate has argued similarly in Bate 1991 and 2000.)

In his discussion of German civilisation criticism, Amery focuses on Catholic writers, ignoring Oswald Spengler, Ludwig Klages and Ernst Jünger. Nor does he concern himself with the influential left-wing critiques of progress written after the Second World War by Friedrich Georg Jünger, Horkheimer and Adorno, and Günther Anders, which provided indirect links between civilisation criticism and the environmental movement in the 1970s. The magazine *Das Labyrinth*, he comments elsewhere (Amery 1991: 39), to which Heinrich Böll and the artist H.A.P. Grieshaber contributed in the early 1960s, was one of the last attempts to expound Christian conservatism in the tradition of this proto-ecological civilisation criticism. Had it lasted longer, it might have served as a bridge to the new social movements in the 1970s. The absence of a movement with the potential to combine the old faith with cultural renewal and to find a natural ally in the Left accounts for the negative reception of Ivan Illich (author of *Tools for Conviviality*: Illich 1973) in Germany, and the scorn poured on *The Limits to Growth* (Meadows et al. 1972) by orthodox Marxists. Only via the international scene has Christian conservative civilisation criticism of German origin exercised any influence on the environmental movement, in the persons of the South American missionary of Austrian Croat origin Illich, the émigré Ernst Friedrich Schumacher, author of *Small is Beautiful* (Schumacher 1973), and Eugene McCarthy, the anti-Vietnam War hero of radical American youth in the 1960s, whose policies were inspired by his education by German Benedictines and G.K. Chesterton's Land Distribution League (Amery 1991: 45–7). These historical perspectives on proto-ecological counter-traditions to progress and economic growth also play a role in the novels *Das Königsprojekt, Die Wallfahrer* and *Das Geheimnis der Krypta*.

Nearly twenty years separate *Natur als Politik* from Amery's third major non-fiction publication on the environment, *Die Botschaft des Jahrtausends* (full title: The Message of the Millennium. Life, Death and Dignity). Here Amery asks why the environmental movement has failed, criticises Green excesses and false sentimentalities, and seeks to establish a reliable system of values and guidelines for a culture of sustainability. Amery's message remains one of active commitment to ecological enlightenment, balancing ecological needs against social justice and humanity, and individual humility and renunciation to the point of asceticism.

Given the dilemmas posed by overpopulation and resource depletion, there is a real danger of political leaders and scientific experts resorting to solutions involving the discrimination, enslavement or extermination of sections of the world population, in order to enable a chosen few to continue to live a life of ease. Nonetheless, Amery regards himself less as a Cassandra prophesying doom than as an Isaiah, Jeremiah or a Jonah, calling for the sinful to repent and be saved. The twin solutions, simple but morally unacceptable, of biological barbarism, as practised by Hitler and Pol Pot, and technocracy, with its subtler means of selection and its penchant for false, high-tech solutions to the problems of transport, energy and food production, must be resisted. We need a new culture investing voluntary renunciation, acceptance of ageing and death, and even suicide, with new dignity.

Amery revises the definition of his adopted position of Deep Ecology to include a scale of values applying in situations demanding a choice between saving human or animal life. We must acknowledge the killing and dying inherent in life, rather than hiding it, for instance in our farms and slaughterhouses, and ask ourselves what forms of promoting and ending life are conducive to the health of the biosphere. Leaving open what the future holds in store, including the possible demise of the species, Amery comes close to Hoimar von Ditfurth's advocacy of stoic acceptance of death and the meaninglessness of human existence in the face of the supreme indifference of Gaia in the bestselling book *So laßt uns denn ein Apfelbäumchen pflanzen* (Then Let Us Plant an Apple Tree: von Ditfurth 1985). However, while von Ditfurth argued that innate, biologically programmed mental structures and patterns of behaviour precluded our changing and controlling the development of civilisation and the human race, Amery remains optimistic. Christian faith is a part of the solution. It may not be essential to endow the simple lifestyle demanded by the future with dignity and meaning, but men and women who have lived lives of contemplation and renunciation can provide role models.

Hitler as a Predecessor

Amery's most recent book, *Hitler als Vorläufer*, expands on an already familiar theme. In addressing the links between Nazism and sustainability, Amery broaches a topic particularly uncomfortable for German Greens. Environmentally interested novelists and poets in the 1970s and 1980s such as Günter Grass, Günter Kunert, Peter Härtling and Wolfgang Hildesheimer had aligned environmentalism with antifascism. They saw in the holocaust the ultimate example of the destructiveness of modern civilisation, exemplifying disregard for human life, natural cycles, dependence on nature, genetic and cultural variety. However, historians of the environmental movement, philosophers and cultural commentators have argued the relationship between ecology and fascism was (and is) more complex.

Anna Bramwell's pioneering history of environmentalism in the twentieth century (Bramwell 1989) appeared to seek to vindicate extreme right-wing politics by arguing environmentalism was an integral part of it. Fortunately, the American historian of German environmentalism Raymond Dominick has since provided a lucid, balanced assessment of the implication of environmentalism in Nazism (Dominick 1992). The doctrine of Blood-and-Soil was a political instrumentation of the social anxieties and environmental concerns encapsulated in the *Heimat* movement and early twentieth-century civilisation criticism, and the Hitler Youth adapted key elements of the *Wandervogel* protest movement. However, despite Hitler's vegetarianism, Darré's interest in organic farming, Todt's exemplary landscaping of the new *Autobahnen* and the forward-looking Reich conservation law, Dominick points out that many of the central principles of today's environmental movement (e.g., the ecological world view, concern for public health, interest in Eastern philosophies, empathy with non-human life and respect for God's creation) are fundamentally incompatible with the Nazi principles of leadership and absolute subordination, nationalism, militarism, male chauvinism and antisemitism. (Thomas Rohkrämer similarly demonstrates the eclecticism of National Socialism in his contribution to this volume, and points to the increasing dominance of technocratic ideas in actual practice over blood-and-soil mysticism as time went on.) Andrew Dobson has pointed out that no one form of society is singularly appropriate to sustainability, and political responses have ranged from conceptions of a 'new global order' and 'centralised authoritarianism' through the 'authoritarian commune' to the 'anarchist solution'. (Dobson 1995: 80–2)[6] Nonetheless, the troubling historical implication of environmental radicalism in totalitarian politics has provided critics of the Greens with welcome ammunition. Richard Herzinger and Hannes Stein have marshalled the arguments against 'Totalitarianism in Green' in their broadside against the culture and politics of the 'post-1968 generation' in Germany, *Endzeit-Propheten* (The Prophets of Doom: Herzinger and Stein 1995: 78–86). The French philosopher

Luc Ferry's book *The New Ecological Order* presents a troubling analysis of the common ground between Fascism and Deep Ecology in their rejection of 'anthropocentrist modernity' and 'relentless hatred towards all forms of humanistic culture – in particular the disgraced heritage of the Enlightenment' (Ferry 1995: xxvii-xxviii). After examining the 1933 Reich Law for the Protection of Animals, the Reich Hunting Law (1934) and the Reich Law for the Protection of Nature (1935), Ferry concludes these no longer positioned man 'as master and possessor of a nature which he humanises and cultivates, but as *responsible* for an original wild state endowed with intrinsic rights, the richness and diversity of which it is his responsibility to preserve forever' (p. 107). He implies physiocentrism, or belief in the intrinsic value of nature, led and leads to inhuman brutality. Susan Bratton has analysed the flaws in this seemingly plausible argument. She points out the physiocentrism encapsulated in the Reich Hunting Law was not typical of practical activity in the Third Reich. Like many other ideologies, Nazism was inconsistent in terms of its philosophy of nature. The privileging of nature over the human, biospherical egalitarianism, or even just intrinsic respect for nature were in fact less central to it than the dualistic division of both the natural and the human spheres into over/under, good/evil and beautiful/ugly. The Nazis' conception of nature was thus a non-inclusive one, excluding whole spheres of reality from the 'natural', the 'living', the 'authentic', the 'organic'. Hence the existence of 'environmental' antisemitism: the Jews were classed as cruel to animals (engaging in ritual slaughter), unnatural and inorganic. Bratton concludes: 'Raising the value of nonhuman nature becomes dangerous primarily when some humans are assumed to be external to or divorced from the natural' (Bratton 1999: 19).

Far from seeking to deny the politically dangerous potential of radical environmentalism or to refute allegations of its proximity with the Nazis' religion of nature, Amery turns these to his advantage. He interprets the horrors of the Third Reich as the product of rational environmental concerns and presents them as a warning to posterity against the temptation to trade human dignity and liberty against personal security. Hitler's euthanasia programme and the holocaust, he argues, obeyed a twisted logic driven by preoccupation with the future survival of humankind. The roots of Nazism are to be found in the nineteenth-century German tradition of civilisation critique and cultural pessimism, which combined legitimate concerns over the scarcity of material resources with a brutal materialistic form of Social Darwinism, encouraging eugenics and scorn for democracy as a weak, merely mechanical system involving constant compromises.

Amery had already mentioned Hitler's vulgar Darwinistic conception of the laws of nature in passing in *Natur als Politik* (Amery 1985a: 218), and devoted most of a chapter to the idea in *Die Botschaft des Jahrtausends* (Amery 1994: 69–88). How would the various political parties in the Weimar Republic have reacted, he asks there, if a work such as *The Limits to Growth* had been written in the late 1920s? The Communists would have dismissed it out of hand, the Social Democrats denounced it as cultural pessimism, and the Christian parties rejected it, confident in the redemption of humankind. Hitler, however, would have embraced it. The Thousand-Year Reich, he argues provocatively, was Hitler's answer to what will be the great challenge of the twenty-first century, the depletion of natural resources and the running out of *Lebensraum* for a growing population. A master race equipped with all the advantages and conveniences of modern technology was to guarantee the sustainability of planetary life, and of human cultural achievement. Hitler's programme sought to postpone the exhaustion of resources indefinitely by aggressive conquest, the enslavement of inferior races, the extermination of superfluous populations, and the replacement of humanist respect for the individual with a ruthless efficiency modelled on the 'aristocratic principle' of nature. The 'natural' way of life Hitler strove for was one of Darwinian natural selection unhindered by sentimental inhibitions. Amery goes on to argue that despite the unlikelihood of such views being taken seriously again in Germany, they possess a fatal attraction for the peoples of the Third World, whose exploitation has continued apace since 1945. The popularity of current concepts of planetary management suggests many people would be prepared to give up human rights and the protection of the underprivileged in order to safeguard civilisation and their own standard of living. Environmental dictatorship remains a serious threat as long as Western society fails to come up with an answer to the conflict between biospherical stability and population growth.

This interpretation of the 'sustainability aspect of Hitler's programme' is expanded and further developed in *Hitler als Vorläufer*. Amery's argument is based on two striking passages in *Mein Kampf*, at the end of Chapter 2 and in Chapter 4, which indicate Hitler was concerned to create a sustainable state and ensure the survival of humankind – led, of course, by Aryans. The first of these envisions an evil triumph of Judaism and Marxism, which would leave the Earth as a depopulated, barren planet encircling the sun:

> If, with the help of his Marxist creed, the Jew is victorious over the other peoples of the world, his crown will be the funeral wreath of humanity and this planet will, as it did millions of years ago, move through the ether devoid of men. Eternal Nature inexorably avenges the infringement of her commands. (Hitler 1935: 60)

'Hence today', he concludes in an infamous phrase, 'I believe that I am acting in accordance with the will of the Almighty Creator: *by defending myself against the Jew, I am fighting for the work of the Lord.*' Hitler, as is well known, saw in the Jews the representatives of pacifism, internationalism, humanism and liberalism, and associated them with both Marxism and Christianity. Here and elsewhere in *Mein Kampf* he links them with concern for the preservation of life, compassion with the handicapped and weak, equality and respect for human dignity. There are four possible ways of dealing with population growth, he argues in Chapter 4, namely birth control, increased agricultural efficiency, expansion into new territories, and the purchasing of foodstuffs from abroad with money earned by industrial production and trade. The second and fourth are inherently problematic. His preferred solution is of course the third, and he believes Germans have a right, as the superior race, to take land from their Slav neighbours. It would be foolish weakness to shy away from the starvation (or violent death) or inferior peoples, and resort to birth control, since this would not allow for natural selection. Eugenics is the only alternative to decadence and doom for humankind. Hitler identifies himself with the great aims of the 'cruel queen' Nature, which demonstrate, in a provocatively paradoxical phrase, 'the *humanity* of Nature, which destroys weakness in order to grant strength its place'. He dismisses other conceptions of humanity as attempts to cheat nature which are doomed to failure.

Despite these passages, there is in fact little in *Mein Kampf* to suggest Hitler was genuinely concerned with *Lebensraum* for the whole of the human race, rather than for the Germans, when he dictated the book in 1923.[7] Amery's presentation of Hitler as an environmentalist is, therefore, less concerned with historical accuracy than with underlining the threat of a new eco-fascist state unless some far-reaching, more democratic form of action is taken. A more serious weakness is that, for all its provocation, *Hitler als Vorläufer* offers few points of engagement for contemporary ecological debate. Amery does not, for instance, explore parallels with the quasi-dictatorial powers of central government in the environmentalist visions of Wolfgang Harich, Herbert Gruhl, Rudolf Bahro, Hans Jonas or the Neonazis (see Herzinger and Stein 1995, Jahn and Wehling 1990, Geden 1999). The principal perspectives for action he himself offers have not changed significantly in twenty-five years: these include, beside voluntary asceticism, the draconian abolition of eighty percent of current industrial production as *Wohlstandsmüll* (the rubbish of affluence, p. 171). No more than Amery's previous books does this probe the problematic nature of radical opposition to modernity and rejection of humanist anthropocentrism when translated into politics.

Like the novels *Der Untergang der Stadt Passau, Die Wallfahrer* and *Das Geheimnis der Krypta*, it could even be held to participate in the (unacknowledged) longing for apocalyptic destruction as an agent of purging and punishing which characterised so much environmentalist discourse in the 1970s and 1980s. It is only fair to say that *Hitler als Vorläufer* probably sought less to contribute to historical research or ecopolitical theory than to cultural criticism. Here too, however, its impact was limited. In the climate of declining public interest in environmental politics it was perhaps inevitable that it would be less widely reviewed than *Das Ende der Vorsehung* or *Natur als Politik*. In the event, it has been generally welcomed as a salutary reminder of the danger of sacrificing democracy and human values in pursuit of sustainability. Scientific progress, clean technology, and the apparatus of legal and economic incentives for environmentally friendly behaviour have yet, after all, to solve the conundrum of ensuring lives worth living for future generations.

Creative Writing and the 'Preservation' of Nature

Looking back at *Das Ende der Vorsehung* and *Natur als Politik* in an afterword written for a reprint in 1985, Amery observed that it had been his express intention to bring about change in the value systems and attitudes of his contemporaries (Amery 1985a: 369). Given the failure of Germany (or any other industrialised country) to make the kind of radical change demanded by the environmental situation in the intervening years, his books, he suggests, might logically be classified as failures. When free copies of *Die Botschaft des Jahrtausends* were sent by the publisher to the 672 members of the *Bundestag* in 1994, only ten of them bothered to acknowledge receipt (Kiermeier-Debre 1996: 196). Nevertheless, as Baukloh and Roose have shown above, there was undoubtedly a significant shift in the direction of environmentalist values in Germany in the 1970s, not all of which has been reversed in the 1990s. It is well-nigh impossible to measure the impact of Amery's essays on environmental debate in Germany, but they seem likely to have contributed towards this shift and helped consolidate it – *Das Ende der Vorsehung* by exploring the cultural, philosophical and religious roots of the environmental crisis and formulating its consequences, *Natur als Politik* by challenging politicians on the Left as well as Right with a theory of ecological materialism, and *Botschaft des Jahrtausends* by reflecting on the reasons for the decline of the environmental movement in the 1990s, and formulating an environmental ethic.

Amery's essays have contributed to an evaluative discourse of moral and aesthetic persuasion rather than scientific elucidation or philosophical deduction. Focusing on the cultural factors determining our interaction with the environment and foregrounding the ethical and religious dilemmas we are increasingly faced with, his strategy is one of exaggeration and polemic rather than balanced argument, providing historical illustrations and quotations from authorities in place of factual information or logical proof. There is more than a touch of religious exhortation in his targeting of indifference, acquiescence, stoicism and cynicism. If, at times, he oversimplifies issues, or slips into cultural pessimism (see Amery 1991: 179 and 1994: 89f.), at others he is capable of writing with infectious humour and Swiftian irony. (See, for instance, Amery 1991: 82, where he proposes an international institute be founded for the decimation of humanity.) His strategies of persuasion include repetition and variation, ironic juxtaposition and the dramatisation of key points in fictional interludes. There has been broad agreement among reviewers that Amery's essays constitute a singularly intelligent, entertaining and thought-provoking commentary on the environmental blind spot of contemporary society.

'What is required of them is less the construction of a logical argument', Carl Amery once said of writers seeking to change society, 'than seeking out the unconscious regions in which people's prejudices lurk, and devising suitable ways of combating them' (Töteberg and Smith 1999: 2). In his own novels, dramatic scenarios of global catastrophe and alternative, sustainable societies and cautionary tales of planetary management, presented with inventiveness, humour and stylistic flair, borrowing from the genres of the *Bildungsroman* (merging individual with cultural and historical processes), detective story, science fiction, the historical novel and travel writing, are among the ways in which he engages, obliquely but no less effectively, with readers' prejudices, anxieties and aspirations. Amery has alternated between fiction and non-fiction, and combined elements of both. He has also alternated and combined in his writing the three principal paradigms of environmentally committed writing identified by Dagmar Lindenpütz in her recent study of children's literature, namely *enlightenment, ethical guidance* and *warnings of catastrophe*. The practicioners of ecological and political enlightenment, Lindenpütz explains, put their trust in the efficacy of knowledge and rational argument, while others seek to provide an ethical foundation for action by example, and others again, perhaps the majority of writers today, present postmodern diagnoses of the age, in the form of ironic scepticism or apocalyptic pessimism. These last constitute alternative responses to the increasing alienation, fragmentation and self-destruction of modern society. All three kinds of writing, Lindenpütz argues, make valid, complementary contributions to the environmental education of young readers. (Lindenpütz 1999: 52f., 241f. See also Lindenpütz's contribution to this volume.)

The task of literature and art, Schiller argues in 'Über naïve und sentimentalische Dichtung' ('On Naïve and Sentimental Writing', a key text of German classical aesthetics first published in 1796: Schiller 1962), is to remind us what nature was like, to preserve it, and to facilitate our return to the lost state of nature on a higher level. Schiller's ideas have been revisited and reformulated in the light of the 'dialectic of the enlightenment' (Horkheimer and Adorno 1973) and the destructive potential of modern technology, first by Adorno (Adorno 1970: 115), and more recently, in an explicitly environmentalist context, by Hartmut and Gernot Böhme (Hartmut Böhme 1988: 30f., 147 and Gernot Böhme 1989: 15), and Klaus Michael Meyer-Abich. The prime function of culture, according to Meyer-Abich, lies in intregrating humankind in nature as a whole. The role of literature and art is to revive the sense of a physical, aesthetic relationship with nature which is fast disappearing in the alienation of modern society. Our dilemma today is that we know the damage we are doing to the environment, but we continue in the old ways, endangering vulnerable peoples and future generations. By furthering appreciation of the beauty and dignity of nature, and recognition of ourselves as a part of it, the writer can contribute to perceiving, feeling this damage. This alone holds out the promise of changing our behaviour. (Meyer-Abich 1990: 12–24, 97–100, 113–7)

Amery subscribes to this holistic aesthetic in the essays and speeches collected in *Bileams Esel* (Amery 1991). He writes, for instance, of the 'potential active role' of art in bringing about a new orientation in thinking and feeling:

> This [new] consciousness must orientate us towards the network of life on the planet, and teach us to conceive of the planet as a self-regulating, self-preserving organism; it must reveal to us the impossibility of continuing to hold by a philosophy which leads at worst to limitless domination of nature, and at best to a partnership of equals between an independent humanity and an independent nature confronting it. Such a reorientation is a challenge for all our faculties, it is a cultural challenge in the widest sense. A religiosity which attempted to circumvent it would be as foolish an undertaking as an aesthetic which ignored it. This is the natural and most pressing challenge facing the art of the present and the future. (Amery 1991: 250f.)

However, he had already stressed in *Das Ende der Vorsehung* that overt political and environmental activism in creative writing is less effective than indirect treatment:

> The anger of the battle cry against the great global injustices is actually less impressive than quiet implacableness on the verge of insanity – Hölderlin, Kafka and Beckett have more to say about the human condition than Brecht or Bloy, whose political engagement would appear so much stronger. (Amery 1985a: 131)

Not 'Umweltdichtung'(writing about environmental problems) but rather 'Welt-Dichtung' (writing about the world) is then called for (Amery 1991: 270). The most important contribution of art and literature to efforts to preserve creation is an indirect one, namely education in an aesthetic appreciation of nature, he argues, like Meyer-Abich, in the essay 'Die wahre Wende' ('The True Turn-around'): 'The atrophying of our aesthetic sense and our artistic creativity has contributed a good deal to the failure of our social alarm systems to respond to social and ecological dangers' (Amery 1991: 144).

Three of Amery's essays are directly concerned with the social and environmental task of the writer: 'Nachrichten aus der wahren Geschichte' ('News from True History'), 'Gedanken über Natur, Naturschutz und Kunst' ('Thoughts on Nature, Conservation and Art') and 'Macht und Ohnmacht der Literatur im Zeitalter der biosphärischen Verantwortung' ('The Potential and Impotence of Literature in the Age of Biospherical Responsibility'). Amery writes of the duty of journalists, artists and novelists to lead the revolt against the myth of technical expertise, as opinion formers and moralists. In the public hearings held in the mid-1970s when a nuclear power station was planned in Wyhl, it became apparent that so-called experts were ignorant of the police surveillance necessary for nuclear safety, the restrictions on personal freedom, and the biological and climatic consequences of accidents. Not to mention the moral and theological dimensions of building power stations whose waste would need to be so carefully guarded for thousands of years. It is the task of the writer 'to introduce news from true history into a public and published opinion dominated by vested interests on the one hand and by anxious fears on the other... as an anti-expert, as a generalist, a trouble-maker who dares to query the emperor's new clothes' (Amery 1991: 246). He is less sanguine regarding the ability of novelists to represent adequately the new environmental consciousness we need. However, this is an area in which poets have made important contributions (p. 268).

Amery's defence of nature is linked with a defence of beauty, individuality, creativity, humanity, and the arts in a world of norms, efficiency and materialism. In a recent newspaper article he identifies as one of the greatest problems facing the Green movement the dirth of works illustrating the existential environmental dilemma of the human race and motivating readers to confront it. We need texts exposing the life-lie of modern civilisation, he argues, and rejuvenating a weak and cynical society:

> What makes the task of the Green movement even more difficult is the striking failure of culture and literature to give shape to the overarching reality, the crucial perspective, in art, literature, the humanities, even (perhaps especially) theology... It's not a question of sketching out yet another utopia, but of liberating us from cherished lies – and, perhaps, rejuvenating a society which has become weak and cynical. (Amery 1996)

In the absence of sociological surveys, the contribution of literature to environmental debate in Germany over the last few decades can only be speculated on. Novels such as the prizewinning *Wenzels Pilz* (Wenzel's Fungus), written by the biogeneticist and ecologist Bernhard Kegel (1997), or Johannes Mario Simmel's (1990) best seller *Im Frühling singt zum letztenmal die Lerche* (The Lark's Last Song in Spring) may have served as Trojan horses, gaining public attention for more or less thoughtful discussion of pressing environmental issues. However, such popular novels operate within the constraints of the genre, and seem likely to have exercised little lasting influence over readers' views, let alone their behaviour. Less carefully written works have fallen into the trap of oversimplification, provoking emotional reactions based on half-understood problems. Carl Amery's novels and essays may, however, have realised some of the potential of literature to inform, motivate and empower readers. By presenting scenarios of environmental crisis they have dramatised the situation confronting us, by constructing counter-images of nature to the resourcist, mechanistic ones which dominate our everyday lives, they have made alternative perspectives more credible. By infusing realism with imagination, they have enriched the lives of readers and invested them with the sense of hope essential to the successful resolution of the problems the future holds in store.

Notes

1. The most detailed account of Amery's life and literary work is Kiermeier-Debre 1996. See also Töteberg and Smith 1999; Hanuschek 1998. Julie Klassen's article '*Natur als Politik*: Carl Amery's Ecological Philosophy' (Morris-Keitel and Niedermeier 2000: 153-165) is principally concerned with Amery as a Green thinker.

2. The *Father Brown* stories are 'not so much ... detective stories ... as parables, in which moral theology is presented as detection' (Stringer 1996: 123).

3. The title echoes the economist Kenneth Boulding's essay 'The Economics of Spaceship Earth' (1966), which contrasted the reckless exploitation, consumption and production of the 'cowboy economy' with a future 'spaceman economy'.

4. In the light of the substitute function detected by feminist theorists in many (male) writers' gendering of nature, either as mother or lover, it is perhaps worth noting that Amery lost his mother, who died giving birth to her second child, only a month after his first birthday. The possible link between this experience and his recurring preoccupation with the female archetype cannot be further explored here.

5. My thanks are due to the German Academic Exchange Service (DAAD) for supporting a study visit to Germany in 1999 which enabled me to consult the Carl Amery archive in the Monacensia library, Munich, and relevant literature in the Deutsches Literaturarchiv, Marbach.

6. Dobson actually goes on to argue that ecologism as a political ideology is necessarily situated within a 'broadly left-emancipatory framework' (p. 85). There is 'some evidence' to suggest democratic institutions are more conducive to sustainable living than authoritarianism: information flows lacking in totalitarian states are necessary for effective policy-making, and authoritarian regimes lack legitimacy in the long term. Links between ecologism and authoritarianism become 'even more implausible' when one considers principles rather than objectives: Dobson cites Robyn Eckersley's view that ecologism is fundamentally emancipatory in its focus on the self-determination of all entities, including humans.

7. Amery does not discuss the source of Hitler's concept of nature in any detail, merely pointing to the political geographers Karl Haushofer and Friedrich Ratzel. Haushofer tutored Hitler and Rudolf Hess in Geopolitics at the time *Mein Kampf* was written in prison at Landsberg, and has been called Hitler's 'intellectual godfather'. He did indeed base his racist theories on Malthusian logic, but can hardly be described as an environmentalist. See Herwig (1999), where *Lebensraum*, Autarky, Panregions, Land Power vs. Sea Power and Frontiers are identified as the five principal concepts in Haushofer's writings.

CHAPTER 10

Environmentalism and its Cultural Transformation in the German Democratic Republic: Poetry and Fictional Prose

Jacquie Hope

As with so many problems of the former German Democratic Republic, the details of the ecological crisis there were only revealed as a result of the *Wende*, the events of autumn 1989 which paved the way for unification. Its scale exceeded the most pessimistic previous estimates, owing to a range of well-known causative factors.

A policy of forced industrialisation had been implemented in response to wartime damage, Soviet reparations, and the deficiencies of a largely rural area cut off from the industrial heartland of the Ruhr. Outdated, heavily polluting plant had remained in commission owing to the sluggish economy. The absence of domestic alternatives had led to reliance on lignite (*Braunkohle*) for energy generation. Moreover, environmental damage had not been effectively mitigated, as in the West, by public protest and debate. Air pollution had reached a level causing widespread health problems; forest dieback was advanced. Many waterways were heavily contaminated, and poisonous drinking water was being supplied to millions. Lignite extraction by means of open-cast mining had devastated vast areas; the rural landscape and ecology had been severely damaged by the extreme implementation of agricultural modernisation. The GDR's atomic power stations were of a particularly dangerous design, and appalling contamination had been caused by uranium mining for the Soviet nuclear programme.

Paradoxically, the GDR was the second country in Europe – after Sweden – to enact a comprehensive environmental law, and the *Landeskulturgesetz* of 1970 was subsequently extended by many specific rulings. However, the standards required by these laws were relatively low, and their implementation very incomplete. Although modest fines were exacted from enterprises which exceeded pollution limits, there was insufficient commitment to improving their ecological record. At national level, efforts expended on environmental improvement were frequently outweighed by the exceptionally unecological organisation of the economy. While much attention was paid to recycling, for example, massive structural inefficiency meant that energy consumption per capita was the third highest in the world, despite the modest standard of living.

Ecological problems were exacerbated by the values espoused by official ideology. The humanist impulse of Marxism, dispensing with traditional reverence for the natural world, had been honed by the industrialising fervour of Leninism to create a positive frenzy of enthusiasm for industry, the reshaping of the natural environment, and quantitative material growth. Industrial policy favoured sectors such as steel, cement and carbochemicals, which were destructive to the environment and entirely unsuited to a country lacking natural resources, while propagandist interest in production increases prevented any serious attempt to balance economic and ecological priorities. This was even made explicit in the wording of environmental legislation, which included 'escape clauses' permitting the achievement of production targets at the expense of ecological standards. In comparison with other Eastern bloc countries, this struggle towards higher growth did achieve a degree of success. However, in relation to the developed capitalist world, living standards remained low: the economic benefits were hardly commensurate to the extreme damage done to the environment.

Debate on ecological issues, as on other sensitive topics, was severely limited by the straitjacket of official defensiveness. Explicit legislation in 1982 classified environmental data as secret – a policy upheld until the last months before the *Wende*. The propaganda function of the mass media largely constrained it to anodyne celebrations of the achievements of government policy. In specialist publications, more constructive debates could be pursued. Scientists and economists discussed environmentally-friendly production techniques and economic models, philosophers revisited Marxist-Leninist principles in the light of ecological insights. However, despite the commitment of those involved, such ideas could not be communicated to a wider audience: popular publications were subject to far greater restriction. The extent of this problem is usefully exemplified by the limitations of the book which was widely regarded as the best popular publication on the subject – *Zurück zur Natur* (Back to Nature – Dörfler and Dörfler 1986) by Marianne and Ernst Dörfler. The authors, both well-informed and environmentally committed scientists, offer a comprehensive exposé of general ecological issues, but their text refers only obliquely to domestic problems. Meanwhile, highly challenging theoretical contributions by East German thinkers such as Wolfgang Harich, Rudolf Bahro and Robert Havemann were only available in the West.

More radical debate was pursued within the uniquely independent institutional framework of the Protestant Church, which had from the 1970s famously accepted its socialist context in exchange for relative internal freedom. Church research centres went beyond theological aspects of the issue, commissioning expert discussion papers and hosting conferences in an attempt to compensate for the limitations of other forums. Citizens' groups could associate themselves with the institutional legitimacy of the Church, and this enabled the emergence of an informal environmental movement, alongside the 'official' forum of the *Gesellschaft für Natur und Umwelt* (Society for Nature and the Environment). However, their activities were severely constrained by the limits of official tolerance. The groups' activities were subjected to surveillance and obstruction, demonstrative action was rarely possible, written material was restrained in tone and circulated only locally. Only in the late 1980s were such limitations briefly transcended, as oppositional activity developed towards the apogee of autumn 1989. Outspoken samizdat publications emerged, and the religious affiliation of the groups became more spurious – the mandatory heading 'internal Church information' was prefaced on the cover of the samizdat journal *Arche Nova* (New Ark) by a laconic 'naturally'. Such documents, produced using primitive duplicating machinery and distributed through informal networks, could achieve only limited circulation. However, they reflected the significance of environmentalism within the wider struggle for political emancipation, a status confirmed by the expression of ecological concern within the literary sphere.

The function of creative literature in ecological debate corresponded to the general development of its role within East German society. The initial enthusiasm of literary figures for the establishment of socialism was expressed in their commitment to propagation of official ideology. This role was gradually superseded in the 1970s and 1980s by a more critical solidarity with the socialist project, and the emergence of literature as a forum for more pluralist debate. Literature appropriated a 'fool's licence' which enabled it to express wider perspectives, and took on a heightened significance in the absence of other forms of public debate.

The ethos characteristic of early East German literature constituted a most unlikely prelude to the acute awareness of environmental deterioration which was later expressed in literary form. Industrial construction and material production were enthusiastically cele-brated in the texts of the 1950s and early 1960s – including those of writers who would later express profound ecological sensitivity.[1]

While the overt concern of these texts was the political aim of building socialism, this political *Aufbau* (construction) often appears synonymous with material construction, the (re)creation of an industrial infrastructure after the destruction left by the war. Simul-taneously, the writers' belief in the social and political progress promised by socialism was imbued with a supreme confidence in the benevolence of technological development.

The ideology of industrialisation dominated two distinct phases of GDR literature. In the early 1950s, the *Aufbauliteratur* emerged as writers turned to the portrayal of contemporary life, moving on from the antifascist themes which had prevailed in the immediate postwar years. The second phase of industrial literature was launched by the *Bitterfelder Weg* (Bitterfeld Initiative) of 1959. This project took its name from the location of the inaugural conference – a city regarded at the time as a flagship of industrial growth, but whose fame in later years came to be based largely on its notorious pollution levels. Within the context of the *Bitterfelder Weg*, factory workers were famously exhorted to participate in literary creativity, and professional writers correspondingly enjoined to participate in industrial work, in order better to portray it in their texts.

The relationship between human society and the natural world is typically portrayed in early GDR literature as a battle, in which the forces of industrialisation must prevail in the interest of social progress. This spirit is reflected with particular enthusiasm in the early texts of Volker Braun, who expressed his practical commitment to the *Aufbau* by working at the site where the huge lignite-process-ing complex of *Schwarze Pumpe* was being constructed. He describes the pioneering struggle to create an industrial infrastructure, portray-ing natural landscapes as potential new building-sites. The poem 'Landwüst' (Wild Land) epitomises this approach in its description of the poet climbing a hill to view a largely rural region, and regard-ing the untouched areas as the building land of the future: '*Wirtsberg/ Panoramic view 360 degrees./* The full angle of the future: so far/ one segment is filled.' (Braun 1979: 86f.)[2]

This preference for industrial rather than natural landscapes is developed in some texts into an explicit aesthetic principle. Heiner Müller's early poem, 'Gedanken über die Schönheit der Landschaft bei einer Fahrt zur Großbaustelle "Schwarze Pumpe"' (Thoughts on the Beauty of the Landscape during a Journey to the Schwarze Pumpe Construction Site – Müller 1958), was inspired by the same site: it records the poet's reaction to the various scenes observed during his journey. Having passed through agricultural landscapes, and observed the labour of traditional farming, he perceives the nascent industrial complex as more beautiful, because it represents social progress and the promise of an easier life. 'The new, more beautiful landscape' is seen embodied in a scene of 'Chimneys. Construction halls. Steel and concrete./ Earth, torn up, mountains, moved with machines and/ noise and dust'.

The heroic view of industry and technology expressed by GDR literature of this period was clearly influenced by early Soviet works: it was somewhat anachronistic in the East German context. This was not, after all, a peasant country in which socialism represented a belated opportunity of modernisation. Johannes R. Becher's poem, 'Volkes Eigen' (Owned by the People – Becher 1959: 119), which celebrated nationalised energy generation in euphoric terms, elicited from a Western critic the sobering observation that 'the lights which went on in the postwar years in the Soviet zone of occupation simultaneously went on in the Western zones, and had already shone during the Third Reich' (Zimmermann 1984: 61). However, the pioneering consciousness of the Soviet writers found a certain parallel in the GDR's situation of existential competition with the capitalist reconstruction in West Germany. Industrial construction is recurrently described using the metaphors of warfare, with invocations of heroism, victories and defeats.

Dissenting voices were few and far between in these early years. Günter Kunert – who famously pursued his technological pessimism both before and after his emigration to the West in 1979 – provided one notable exception, with the brief lyric masterpiece 'Laika' (1963, in Kunert 1969: 54) and his contribution to the well-known 1966 debate in the youth magazine, *Forum* (Bahro et al. 1966: 23).[3]

In the 1970s and particularly the 1980s, however, as the relative cultural liberalisation of the Honecker years coincided with international growth of environmental awareness, creative literature became a rare haven for published ecological expression in the GDR. In lyric poetry in particular, the ethical imperative of reshaping the landscape was replaced by an emphatic identification with the beleaguered natural world. Dying trees, polluted waters, and landscapes despoiled by mining and agricultural modernisation are portrayed in texts whose critical force even in retrospect appears overwhelming. In prose, ecological sensitivity took its place alongside other themes expressing the general reorientation towards subjectivity – away from the aggressively materialist emphasis of the early years – and the increasing distance from officially-espoused ideological positions. Environmental issues were thematised, and broader critiques undertaken of the values of contemporary industrial society.[4]

Literary figures also expressed their ecological commitment through practical engagement. From the early 1980s, a series of annual gatherings known as the *Brodowiner Gespräche* (Brodowin Discussions) brought together interested writers to discuss the issues and share information.[5] At the Tenth Writers' Congress in 1988, environmental debate emerged into a considerably more public forum, as controversial speeches were made at a gathering attended by the Minister for the Environment, by Erich Honecker himself, and by the Western media (whose broadcasts were widely received in the GDR). Following the congress, approval was finally given for an ecological discussion group within the Writers' Union – an initiative which survived unification and continues in the Berlin-Brandenburg region as probably the only such group in Germany today. The apogee of environmental protest by East German writers was reached in 1989, when the Sorbian writer Jurij Koch initiated an official appeal by the Writers' Union against the destruction of Sorbian villages for open-cast mining (Brezan et al. 1989).[6]

Discussions since the *Wende* on the literature of the GDR have cast a more critical light on the continued fundamental solidarity of the literary establishment with the socialist project, and the limitations of their expressions of criticism. However, a retrospective assessment of literature with ecological relevance still confirms its significance as *Ersatzöffentlichkeit*, an alternative public sphere where controversial ideas could to some extent be expressed. A comparison with non-literary publications – such as *Zurück zur Natur*, mentioned earlier – reveals a striking contrast. More crucial for retrospective assessments of these texts is whether they remain of interest for their artistic merit, or whether their function as *Ersatzöffentlichkeit* appears to predominate. The following discussion of a selection of significant prose and poetry will thus focus on the extent to which their interest transcends their ephemeral value as contributions to an historical debate.

The name with which environmental expression in GDR literature was long considered virtually synonymous was that of the provincial poet and diarist Hanns Cibulka: the international attention attracted by his otherwise unremarkable texts epitomises in both positive and negative respects the *Ersatzöffentlichkeit* function of literature in the area of ecological debate.[7]

The fictional diary *Swantow* (Cibulka 1982) immediately attracted attention in both East and West Germany: its author was

hailed by one Western critic, though with less accuracy than journalistic effect, as 'the first Green in the GDR' (Losik 1981). Both
the ecological and philosophical implications of the text were
indeed remarkable in the East German context. The environmental comments of Cibulka's diarist were widely quoted. Some arise
from the context of his surroundings, in the hamlet of Swantow on
the Baltic island of Rügen: he watches local fishermen catching
deformed, cancerous fish, and observes with foreboding the uncan-
nily silent buildings of the atomic power station at Greifswald. Such
direct experiences provide the pretext for more universal statements, of extreme radicality, amounting to a perspective of
ecological catastrophism.

Similar reflections are pursued in *Seedorn* (Sea-Thorn – Cibulka
1985) and *Wegscheide* (Crossroads – Cibulka 1988), whose narrators
also document their retreat to remote areas for a period of contemplation. The topos of the solitary sage bemoaning a corrupt society
finds new relevance in relation to ecological deterioration, for their
rural surroundings confront each narrator with the direct consequences of contemporary materialism. *Seedorn* is also set on a Baltic
island, while *Wegscheide* describes the Thuringian forest, where the
trees are damaged by air pollution and the berries and mushrooms
poisoned by insecticides.

Cibulka uses the diary form not only to thematise environmental problems, but also to propose potential solutions, conceived at
the level of social values and individual responsibility. His narrators' reflections reiterate the urgent need for reorientation from
material consumption towards inner fulfilment. The orientation of
Cibulka's writing is striking for its fundamental dissonance with
official ideology. The emphasis on individual contemplation rather
than collective action, and spiritual rather than political goals, was
obviously alien to a Marxist approach – and was indeed famously
condemned for this reason in the *Deutsche Zeitschrift für Philosophie*
(German Journal of Philosophy) (Ley 1982). More remarkably,
Cibulka on occasion explicitly acknowledges his indebtedness to
the debates of the Western New Age movement – invoking, for
example, Rupert Sheldrake and his contributions to a post-mecha-
nistic biology (Cibulka 1988: 113–115). Cibulka and his work did
not only attract the attention of Western journalists and the irrita-
tion of East German technocrats and cultural conservatives, but
were significant as a focus of discussion within the unofficial ecological groups and in the wider public. Reviews appeared in the
Church press, and the author's readings throughout the country
provided the pretext for impassioned debates. Outside the GDR
context, the texts appear derivative and didactic. However,
Cibulka's broad influence was perhaps even facilitated by their

unchallenging aesthetic character, and by the discursive quality which renders them usefully accessible and unambiguous.

Lia Pirskawetz's *Der stille Grund* (The Peaceful Valley – Pirskawetz 1985) similarly does not disguise its didactic intent, though the fictional integration is less spurious. The novel deals with environmental problems in a more immediate context, and unlike Cibulka, the author contextualises her reflexions within an invoked framework of socialist values. The narrative centres on a protagonist recently promoted into a position of influence, and on the development of her character as she deals with a conflict of interests affecting her community. However, the nature of the dilemma facing her is deliberately schematic, representing universal conflicts between conservation and economic development, and offering potential reconciliations.

The central character, Carola Witt, has become Secretary of the Town Council in the fictional town of Lachsbach in a rural, mountainous region of the GDR. She is faced with the difficult task of judging a planning application from the local factory, which wants to build new premises in an area of natural beauty close to the town.

The text uses this conflict to provocatively examine the uncritical commitment to quantitative economic growth which appeared to condemn 'actually existing socialism' to becoming a rather unsuccessful imitation of capitalism rather than any truly radical alternative. The factory's managers assume that the projected production increases will be given absolute priority over other considerations – even though this is technically illegal, because it contravenes the landscape protection laws. A dream journey into the town's history audaciously makes explicit the comparison with capitalism: it emerges that the industrialists who founded the factory in the nineteenth century had very similar values.

Moreover, the didactic intent of the text goes beyond criticism of the status quo and seeks to propose alternative political values. The true dilemma perceived by Carola is between economic and environmental aspects of social welfare – between job creation and improved working conditions, and the recreational and spiritual value of the beautiful landscape. The prevailing model of socialism favours crude quantitative measures of progress which make the latter difficult to defend. The text engages in a quest for possible alternatives, the decision-making responsibility of the central character providing the pretext for an exhaustive, though ironised, exposé of perspectives from the relevant academic literature. Despite its fictional integration, this material is clearly incongruous within a literary work, and reflects the author's strong publicistic motivation, made explicit in other statements (Pirskawetz 1987).

The key to the *Ersatzöffentlichkeit* function of Monika Maron's *Flugasche* (Flight of Ashes – Maron 1981) also lies in the professional position of the central character. Josefa Nadler is a journalist, as was Maron before she became a full-time writer. The stimulus for the writing of *Flugasche* was as a direct alternative to journalistic expression – to reflect on an environmental scandal which the author had been unable to expose in the East German media. In the end, the novel itself also fell victim to censorship and was only published in the Federal Republic.[8] However, in the paradoxical situation of the GDR, where restrictive cultural policy was radically undermined by the ubiquitous reception of the West German broadcast media, the attention provoked in the West by its censorship led to widespread awareness of the novel in the GDR.

Like Maron, Josefa is sent to report on the city of Bitterfeld (spuriously disguised as 'B.'). Instead of composing the anticipated eulogy about heroic workers and social welfare achievements, she finds nothing more urgent to write about than the appalling air pollution. The novel enumerates the facts which should have been published in journalistic form: B. is the dirtiest town in Europe; 180 tons of airborne ash fall upon it each day. It reveals the distorted political priorities which the journalist wishes to expose: a cleaner power plant is under construction, but the outdated one which causes the pollution will not be decommissioned – a marginal economic gain is given priority over a major environmental improvement. It juxtaposes the social progress achieved under socialism with the ecological cost which undermines it: the children benefit from free health-care and holidays at the seaside, but many have environmentally-induced asthma.

In the case of *Flugasche*, despite the author's directly publicistic motivation, the novel's artistic success is not obviously compromised by its political role. The thematisation of journalism creates a natural unity between the material and the polemical aim of the novel. The text portrays pollution through the eyes of a reporter, but in a manner equally successful as realistic literary narration: through its effect on the everyday detail of human life. Josefa observes the inhabitants' apparent smile, as they screw up their eyes to keep out the dust; she empathises with the children's health problems, and imagines the burden of cleaning imposed by the ubiquitous dirt. Her situation as a journalist in the GDR means that the dynamic of the plot naturally relates her personal development to the political circumstances which perpetuated ecological deterioration; and the paradox of writing about censorship is elegantly reflected by the narrative perspective, with its predominance of internal monologue and imagined dialogue.

The more problematic aesthetic aspects of the novel can also be perceived as implicitly relevant to the ecological crisis. The text manifests an apparent disjunction between its two major themes – the outer events relating to the Bitterfeld issue, and the inner crisis of the protagonist. However, at a deeper level – which the text makes only tentatively explicit – the two appear philosophically as well as circumstantially related. Josefa herself perceives her tormented and alienated existence as a function not only of political repression, but of a more general malaise of modern industrial society (e.g., pp. 7–8 and 234–235). Her naïve-romantic perspective is ironised, and yet essentially substantiated by the overall force of the text, in which the dominant image, the unmitigated bleakness of the polluted city, seems paradigmatic not only of an oppressive society, but of a profoundly disharmonious civilisation.

Störfall (Accident – Wolf 1987), the quasi-autobiographical text in which Christa Wolf responded to the nuclear disaster at Chernobyl, occupies a distinctive position within the debate owing to the particular prominence of its author. The effects of this context on the work itself, however, are ambivalent.

Notwithstanding the post-Wende controversies regarding the limitations of Wolf's criticism of conditions in the GDR, *Störfall* is a powerful example both of the importance of her position, and of the resonance achieved by the critical views she did express. The GDR's own espousal of nuclear power made radical comment in the media on the Chernobyl disaster impossible. Only creative literature, and only a writer of considerable stature, could appropriate the licence to speak out. In fact, the relative cultural liberalisation of the GDR's final years meant that debate in literary guise was in this case positively encouraged – the book was published much faster than was customary, in order to facilitate its role as a stimulus for discussion, and Wolf read from the text on East German television. The unique significance of her position is epitomised by the fact that in the popular scientific journal *Spektrum*, debate on the implications of Chernobyl was published only in the guise of response to *Störfall.*

The critical power of this work is wholly remarkable, particularly since its fictional disguise is fairly perfunctory. The narrator concurs with the most alarmist international assessments of the danger represented by the Chernobyl accident – a perspective widely divergent from that of the East German media, which endeavoured to downplay its consequences. She concludes that atomic power generation must be abandoned in view of its obvious risks. Most provocatively, her vehement condemnation of the élites who perpetuate dangerous technologies represents an astonishingly direct statement of political alienation.[9]

In the sheer explicit intensity of its criticism, *Störfall* is in a different league from all other ecologically-aware GDR texts. However, its topicality and critical force were bought at a price in both analytical and literary quality. Both aspects appear unfinished – a quality justified rather problematically by the fiction of the narrator's immediate reaction to the disaster. Political comment remains at a level of incoherent anger towards scarcely-defined authorities, suggesting the frustration of the powerless victim rather than a considered perspective. At the philosophical level, the text muses in a rather derivative and superficial manner on theories of cerebral evolution in relation to the dangerous orientation of science. Stylistically, it reflects the narrator's confusion by self-consciously eschewing elegance, and thus abdicates the opportunity of literary transcendence.

The acknowledged responsibility of *Ersatzöffentlichkeit* is intriguingly apparent from the final page of *Störfall*. The dates of the text's composition are given, and they indicate a much more rapid process than Wolf's otherwise slow and considered mode of writing. This note has the appearance, if not exactly of an authorial apology, at least of an invitation to read the text as a contribution to an immediate debate, rather than as a completed work of art. The narrator's own comments reinforce this: in direct anticipation of the post-*Wende* controversies, she reproaches herself for failing to speak out in the past: 'We have said too little, and said it too timidly and too late' (p. 68).

In contrast, the publicistic function of Irmtraud Morgner's *Amanda: ein Hexenroman* (Amanda: a Novel of Witches – Morgner 1983) is perceptible only from a handful of slightly uncharacteristic passages within a work particularly ambitious in variety and scale. This sequel to Morgner's earlier masterpiece, *Leben und Abenteuer der Trobadora Beatriz* (The Life and Adventures of Beatriz the Troubadour – Morgner 1974) was by far the most successful prose response in GDR literature to the challenge of thematising ecological crisis.

Explicit mention of environmental problems occupies a very minor part of this text. Its focus is the global disruption of ecosystems, a rather abstract concern which is effectively integrated into the fantastic fictional context through the travels of a mythical bird, flying round the world on 'seven-league wings' and reporting on the devastation of the rainforests, the pollution of the oceans, and the dangers of resource depletion and population growth. The factual detail incorporated in these passages suggests an author unable to rely on a reading public well-informed by other sources about current environmental concerns. However, this occasional stylistic dissonance is scarcely disruptive within a text whose unity comprises a mosaic of incongruous narrative strands. These explicit references to ecological issues serve only to establish as the novel's central reference point the endangered state of the contemporary world, both by environmental deterioration, and by the nuclear arms race, which was so central to public concern at the time of the text's composition.

The novel's predominant approach to these issues is not to portray them directly, but to pursue a fantastic narrative, on an appropriately epochal scale, which embodies a philosophical response to contemporary global threats. Adopting the radical feminist premise that such problems result from the distorted social and philosophical structures of patriarchy, the text portrays the archetypes and contemporary embodiments of womanhood as offering the potential for salvation.[10] Subversion, rather than direct opposition, is the favoured tactic, in keeping with the critique of dualism and confrontation; this approach is reflected in the style of the text, which crucially transcends a problem-oriented mode and seeks to embody alternatives.

The three distinct narrative strands pursue this quest in highly contrasting contexts. The material of classical mythology creates the story of a siren reincarnated in the contemporary world, charged by earth goddess Gaia with subverting the destructive orientation of Promethean Man. Nordic mythology provides the imagery of the second plot, in which the eponymous witches, symbolic of the demonisation of the feminine, attempt to overthrow the dominant archetypes of patriarchal cosmology. Finally, a contemporary storyline depicts the challenges of transposing this mythical process into the mundane context of late twentieth-century society. Laura Salman, single mother and train driver in East Berlin, represents the contemporary woman struggling to regain the archetypal power from which patriarchy has alienated her.

The vision of a gentler and more playful world is embodied in the nature of the oppositional activities portrayed. The female characters of the contemporary plot seek to transcend the limitations of dominant rationalism by offering their services as professional fools. At the mythical level of the text, the witches hold a non-violent auto-da-fé where God the Father, Satan, Prometheus, and other venerable symbols of patriarchal cosmology are ritually 'laughed to death'.

Its audacious emphasis on humour gives this text a unique status among GDR prose responses to environmental problems. Morgner regarded this strategy as the only way to portray global crisis in an empowering manner: 'Very serious issues or global situations can only be dealt with using humour. Because they overwhelm. And the human self-preservation instinct rebels against that' (Kaufmann 1984: 1513).

While *Amanda* achieves its distinctive thematisation of ecological crisis by avoiding direct treatment of the issue, much of the poetry which addresses the subject with notable success takes the opposite approach. The powerful lyric evocations of the despoiled modern landscape were the most graphic documents of environmental damage to be published in the GDR in any medium, and many are simultaneously literary work of lasting value. The four texts discussed below represent particularly successful examples of contrasting approaches, and thus give some indication of the variety of aesthetic means with which the production of mere *Ökolyrik* was decisively transcended.

Wulf Kirsten's 'das haus im acker' (the house in the field – Kirsten 1986: 106–8) derives its expressive power from the intense personal emotion of its portrayal of the changing countryside in his native Saxony. It chooses as its theme an object of the destructive effects of the industrial society: an historic farmhouse, due to be demolished in order to create an unbroken expanse of land for modern farming techniques.

The text describes not the new landscape, the proud achievement of socialist agricultural modernisation, but the destruction of the old; it betrays no interest in the increased productivity achieved through such techniques, but laments the loss of the enchanted world of childhood and of the creative inspiration offered by the natural environment. The condemned house, despairingly described as 'a pile of bricks without a future', becomes a paradigm for all the beautiful natural and traditional features of the landscape destroyed by agricultural modernisation.

Though entirely eschewing explicit analysis, the text also expresses a powerful political comment through its evocation of the poet's reaction to the changes. The agents of this environmental manipulation are described as 'megalomaniac princes of the steppes': the ostensibly humane and progressive motivation of socialist agricultural modernisation is dismissed as a pretext for the time-honoured lust for power and personal aggrandisement. The ideological concept of common ownership of the land is implicitly rejected in favour of personal, visceral attachment to the local environment, a naïve identification with the natural world. Indeed, the direct association of the self with the native landscape lends to its reshaping an impression of personal violation. The poet now feels excluded from the land: 'all footpaths/ to paradise only preserved in memory./ the realm of childhood become pathless.' The sense of ownership is, paradoxically and provocatively, associated with the past, since its objects have been destroyed: 'my avenue of cherry-trees –/ heedlessly thrown away, ... my springs/ poisoned... what once/ belonged to me as the air to the bird/ and the water to the fish.'

In contrast, Kito Lorenc's 'Dorfbegräbnis' (Village Burial, 1979 – Lorenc 1984: 50) achieves a strangely haunting evocation of the destruction of the traditional landscape through a depersonalised portrayal. However, the directness of its critique is affirmed by the poem's dedication to the former inhabitants of a specific Sorbian village.

The poem describes with detached fascination the demolition of the village in preparation for open-cast mining. Rather than dwelling on the brutal detail of such destruction, the text evokes an abstract, timeless scene of mythical proportions. The excavator becomes a giant creature in whose jaws the village church is carried away; the lignite is represented as honey, archetypal attribute of the promised land; the image of a honeycomb evokes the high-rise flats for which the mine will provide heating. Sorbian culture is invoked to heighten the anthropomorphic metaphor of burial: wailing women clad in traditional mourning clothes; the funeral urn from which the soul of the village is heard departing. The profundity of individual despair is evoked in spiritual terms through the elderly villager who blasphemously prays for death to end his suffering, in a classical contradiction of socialist optimism. The prosaic mundanity of contemporary society, represented by the drilling brigade and the officer for the preservation of historic monuments, is accentuated by the implicit contrast with this rich imagery and cultural depth.

Heinz Czechowski's 'Landschaftsschutzgebiet' (Landscape Conservation Area – Czechowski 1981: 76f.) employs a narrative approach to convey the effect of the changing countryside on individual experience. The tension between title and text reflects the discrepancy between rhetoric and reality in the socialist management of the environment.

The poet and his friend, seeking to enjoy the natural world by going out fishing, discover that the familiar ponds they had intended to visit are being transformed into an industrial fish farm. This process, ideologically conceived as the appropriation of the land to benefit the community, is experienced as forcible exclusion. The paradigmatic quality of the anecdote is made explicit by the use of political terminology: the poet's way is barred by a group of ironically-termed 'heroes of the socialist reshaping of the landscape', and the text invokes the 'planners and managers' who are directing these developments at national level.

This reshaping of the landscape is portrayed as the creation of a barren wasteland from an area which once embodied the vitality of Nature. The poet evokes the harshness of the steel and concrete used to create the artificial lake, and the violence of the reconstruction process. He invokes the remembered features of the area which are now destroyed, and the impoverishment of the local ecosystems: there is 'No room any more for stork, frog, adder and otter'.

The text transcends a facile identification with the disappointed fishermen as victims of the modernisation process. Having been reduced to purchasing a carp, which despite their protests is wrapped and presented to them live, they experience on their homeward journey 'without knowing why,/ a bad conscience'. Their sense of unease, as they carry home the slowly expiring fish, invokes the unwitting complicity of society as a whole in the processes leading to the destruction of the natural environment.

With his poem 'Elbabend' (Evening by the Elbe, 1987 – Pietraß 1990: 12f.), Richard Pietraß achieves the remarkable combination of a beautiful, evocative, and amusing text which takes as its theme a river polluted by sewage.[11] It describes a walk along the Elbe – ironically, in an important nature conservation area inhabited by rare bird and animal species. The good-humoured self-irony begins in the opening echo of Goethe's *Iphigenie*, 'Seeking with the soul the land of the Greeks': the poet and his companion are described on their evening walk as 'Seeking with our boots the castle of the beaver'. The text laconically integrates its description of pollution with its evocation of the beauty of the river at dusk: while the water reflects the evening light, the sewage debris drifts peacefully by. The sewage is provocatively portrayed as fulfilling a function within a natural world adapted to human carelessness: it provides food for the fish. The text moves seamlessly from a description of animal calls breaking the silence, which inspire the poet to invoke the music of the spheres, to the mention of a floating condom: but the latter is integrated positively into the imaginative framework – it evokes the pleasure of its erstwhile users, and thus provides an appropriate counterpart to the mating 'dance' of the midges around it.

The text does not lament or rail against the ugly human influence on the landscape, but portrays it in a spirit of acceptance of the absurdity of life. The relaxed gesture of the poem, in its lack of protest or exhortation, is nevertheless highly provocative of reflection on the state of the natural world.

The effects of German unification on the environmental situation in the GDR, and on environmentally sensitive literature, have been complex and contradictory. At a pragmatic level, the worst abuses of socialist mismanagement have been ended, and significant resources invested in cleaning up one of Europe's ecological disaster areas. However, as with other aspects of unification, the overwhelming scale of the problems continues to defy any attempt at rapid resolution. Meanwhile, the rapid capitalist takeover has brought new challenges, notably in the pressure on land for industrial, transport and retail developments. In the public consciousness, the challenges of economic insecurity have understandably tended to eclipse the previous prominence of environmental concern. However, ecological awareness in the literary sphere has survived unification, since it obviously transcends the socialist context. Indeed, one prominent post-*Wende* contribution makes this schematically clear: in Volker Braun's *Bodenloser Satz* (Braun 1990),[12] East and West German settings are juxtaposed, in separate sections of the text. Thus, beyond its specific publicistic function within the GDR, ecologically-aware literature can be seen as a clear example of convergence, embodying a shift in the direction of 'post-industrial' values which significantly transcended the East-West divide even before the advent of political unification.

Notes

1. Later texts of both authors cited here, Volker Braun and Heiner Müller, reflect an intensely critical perspective, and many of the writers cited below as representatives of ecologically sensitive writing were at this early stage assertive in their enthusiasm for industry and technology. Examples include poems by Cibulka and Czechowski singing the praises respectively of nuclear power and of the chemical industry, and the euphoria occasioned by the news of the first manned space flight in Christa Wolf's *Der geteilte Himmel* (The Divided Heaven / Wolf 1963).

2. Written between 1969 and 1973, the poem represents a relatively late expression of this ethos.

3. Laika was the dog sent into orbit in a Soviet satellite: her death is provocatively used as a metaphor of the dangers of technology. The *Forum* debate, in which Kunert's views were condemned by none other than Rudolf Bahro, is particularly interesting in view of the latter's subsequent development.

4. This development is revealingly contextualised by Wolfgang Emmerich's account in the 1996 edition of his Concise Literary History of the GDR (Emmerich 1996), in which the chapter on the period 1971–1989 is entitled 'The growing gap between utopia and history: Literature as civilisation critique' (239–395); Axel Goodbody's 'Literature on the environment in the GDR. Ecological activism and the aesthetics of literary protest' (Goodbody 1997) is similarly wide-ranging and insightful. The contributions of anthologies such as Rost (1991) and Herzberg (1991) reflect the breadth of significance of the theme, as does the range of GDR topics in the recent collection of essays on Literature and Ecology edited by Axel Goodbody (Goodbody 1998). Poetry expressing environmental sensitivity was published by a wide range of authors: significant names in addition to those discussed later in this chapter include Volker Braun, Sarah Kirsch, Karl Mickel, Axel Schulze, Jürgen Rennert, Jens Gerlach and Heinz Kahlau. In prose, the tendency towards broader critiques of industrial society makes the theme more difficult to define. Alongside the authors discussed here, significant texts were published by Jurij Koch, Helga Königsdorf, Gabriele Eckart, Armin Müller, Joachim Walther, Joachim Nowotny, Marianne Bruns and Gerd Bieker. The scope of the present study precludes a consideration of drama, but this genre also produced texts expressing both ecological sensitivity and *Zivilisationskritik*, of which Heiner Müller was the most prominent author. The environment was also a theme of children's literature, with Wolf Spillner its most important representative. (See the importance attached to GDR children's literature by Dagmar Lindenpütz in her chapter in this volume.)

5. Brodowin is a hamlet near the Polish border, home of the initiator and initial host Reimar Gilsenbach, author of children's books on environmental issues. (Appropriately enough, it has since unification, on Gilsenbach's initiative, become an 'eco-village', specialising in organic farming.) Gilsenbach has documented the surveillance to which his activities were subjected by the Stasi in his *Trostlied für Mäuse* (Song of Comfort for Mice – Gilsenbach 1994).

6. The small Slavonic group of the Sorbs were the only significant ethnic minority in the GDR: official propaganda emphasised their generous treatment, but lignite mining devastated significant areas of their native region.

7. Anita Mallinckrodt's detailed account of the resonance of this text in its political context (Mallinckrodt 1987) is still useful.

8. The manuscript was accepted by the Greifenverlag, Rudolstadt, but objections from the Ministry of Culture prevented its appearance. A planned edition in 1988 was withdrawn owing to subsequent controversies regarding the author; a renewed promise of publication in October 1989 was rendered irrelevant by the events of the *Wende*.

9. One striking omission, in retrospect, is the contamination caused by uranium mining in the GDR itself: the sensitivity of this operation makes it inconceivable that it could have been a target of criticism, but details of its consequences did not, in any case, come to light until after the *Wende*.

10. This radical feminist perspective, though significant internationally, found no expression in the GDR outside the literary sphere (where it famously also formed the basis of Wolf's novel *Kassandra*, published in the same year, though the latter focuses only on military and not ecological threats).

11. The text is dedicated to Ernst Dörfler, co-author of *Zurück zur Natur*, who introduced the poet to this landscape – a fact directly reflecting not only the close contact between creative writers and other environmentally-active figures, but also the enhanced latitude of expression available to the poet as opposed to the scientist.

12. The word-play in the title renders it untranslatable: 'Boden' is ground, and 'bodenlos' either 'bottomless' or 'incredible'; 'Satz' can mean sediment, sentence, thesis, or leap.

CHAPTER 11

Green Strands on the Silver Screen? Heimat and Environment in the German Cinema

Rachel Palfreyman

It is really not hard to think of German feature films that deal with serious issues of public debate. History, for example, has frequently been a concern of German film-makers: Rainer Werner Fassbinder's so-called 'FRG trilogy', Margarethe von Trotta's *Die bleierne Zeit* (The German Sisters), Alexander Kluge's *Die Patriotin* (The Patriot), Jürgen Syberberg's *Hitler, ein Film aus Deutschland* (Our Hitler) spring instantly to mind. In the light of this, it is perhaps surprising that it is much harder to come up with a similar list of films which deal with the debates and the struggles that have taken place in the Federal Republic over the environment, despite the well-documented commitment to such issues (Riordan 1997b, Kolinsky, ed. 1989, Papadakis 1984). This is due in part to the close connection in film between themes and motifs which examine human interaction with the natural environment and the Heimat complex. Films dealing with environmental themes have often done so via the politically disputed Heimat genre, and no specifically green genre has emerged in feature films.[1] The political connotations of Heimat also meant that even in its critical form, this tainted genre had a rather limited appeal. As Bahlinger, Hellmuth and Reister put it, 'It seems that the directors themselves had problems with the Heimat concept. Nobody – except Reitz – actually intended to make a Heimat film as such.' (1989: 147). So in the 1970s and 1980s when environmental issues were at the centre of political debates in West Germany the (critical) Heimat film was a moderately successful vehicle for exploring environmental questions, yet made it difficult for a more direct genre to emerge. In addition, from the mid to late 1980s an appetite for comedy meant that at the very height of public interest and concern for the environment, such issues were not obvious material for screenwriters. This in turn may be at least partly due to the nature of environmental activism in Germany, which has often grown out of grass-roots projects and regional campaigns, not perhaps lending itself to a large-scale fictional filmic treatment.[2]

In this chapter I shall examine approaches to broadly environmental issues in feature films which allude to the Heimat genre. I shall first discuss briefly how the concept of Heimat is connected with attitudes to the environment in the German-speaking world. I then discuss the construction of the physical environment in the 1930s Bergfilm (mountain film) and in the 1950s Heimat film. Following this, I shall look at the way canonical directors like Werner Herzog and Edgar Reitz appropriated the discourse of Heimat in the 1970s and 1980s to explore the interaction of human beings with their environment in a rather more critical mode. Finally I argue that the influence of the Heimat mode is still palpable in the 1990s and that this influence on constructions of the environment in film has meant that there is no neat fit between modern green politics and the critical Heimat film, but that a diverse range of competing discourses and influences from apocalyptic visions to a Romantic legacy to a utopia of clean technology inform both narratives and images.

Heimat and Environment

The Heimat movement, which gained popularity around the turn of the century, initially encompassed trivial literature, Heimat journals, and local societies seeking to protect their local area and promote the expression of Heimat identity. While essayists such as Adolf Bartels, Friedrich Lienhard and Julius Langbehn tended to take a reactionary line on the moral and spiritual superiority of the province over the degenerate city, and deplored the influence of, variously, modernity, racial alterity, or racial mixing, local Heimat societies had some rather more palatable concerns such as environmental protest against the more disfiguring effects of modernity in their vicinity.[3]

The concept of Heimat implies a sense of identity, roots and belonging. In historical terms, Germany's late unification meant that Heimat discourse functioned as a focus for identity which could negotiate between local, regional and national loyalties. At the centre of the Heimat complex is, then, the relationship of human beings to their physical surroundings. Heimat resides in the invest-ment made by people in the spaces where they live and work, especially where their work depends on the natural environment. Therefore it is easy to see both why Heimat societies became repos-itories for proto-environmental activism, but also why Heimat discourse could be appropriated for such dubious ends as a justifi-cation of colonialism, with the suggestion that land belonged to those who forged a relationship with it through their work.

1930s Mountain Films and 1950s Heimat Films

A popular and influential film genre, the mountain film emerged in the 1920s from the alpine documentary tradition. The genre was pioneered by Arnold Fanck, a director of such documentaries who turned to feature films, and later taken up with equal success by his two protégés, Luis Trenker and Leni Riefenstahl in the 1930s. Trenker continued to make mountain films and to publish illustrated accounts of his mountain career well after the war (Trenker 1961). For many urban cinema spectators in the Weimar Republic, moun-tain films represented their sole opportunity to experience, albeit vicariously, the extreme environments of the Alps. Indeed the mountain film is partly responsible for the popularisation of moun-tain regions as recreational spaces. The innovations in location filming in the face of very difficult conditions are certainly not to be underestimated and while there is still debate about the political symbolism of the mountain film, there is widespread recognition for the cinematic developments Fanck and others achieved, which represent a major contribution to the otherwise largely studio-bound Weimar film industry (see Rapp 1997, Bechtold-Comforty et al. 1989: 43-53, Amann et al. 1992).

Creating a narrative to supplement spectacular cinematic shots of Alpine mountains normally meant that a climbing disaster had to feature, followed by a daring rescue. In these films therefore in their purest form there is often an emphasis on the conquest of a beautiful but dangerous environment, with the male hero pitted against the vagaries of the mountains and nature. The climbers are clearly differentiated from ordinary folk who live in harmony with their rural Heimat, farming the land at the foot of the mountain. Climbers are often characterised as courageous but driven; they break out of the Heimat contract of harmony and seek to conquer Nature. The mountain environment exists as a challenge to their masculine power. This is not to suggest that they are always repre-sented negatively, nor is the rural Heimat shown as repressive. It is, however, often figured as a feminine domain, the hearth to which men might return, but from whence they are also challenged to strike out.

Das blaue Licht (The Blue Light, dir. Leni Riefenstahl 1932) is one of the clearest examples of the environment divided into mediated, cultivated Heimat and pure, wild Nature. The village scenes show a traditional Heimat lifestyle, where the only threat to the rural community is a mysterious blue light from the mountain which appears to call to young men like a Siren to leave their homes and climb a sheer face to find the source of the light. They inevitably fail and fall to their deaths. Junta, an outsider and 'wild woman' played by Riefenstahl, is blamed for this, since she is thought to be a witch. However, a visiting artist finally discovers her secret route to a mountain chamber where a wonderful natural treasure-trove of crystals reflecting moonlight creates the seductive light. However, when he reveals his discovery the villagers arrive en masse to take as many crystals as they can. Their wealth from the exploitation of the natural riches of the mountain is achieved at the expense not only of the mountain itself: it also destroys Junta. Distraught at the sight of the disfigured chamber, she falls to her death. The mountain is here clearly identified with Junta: its rape is equated with an attack on her body and leaves her unable to carry on living. The specific prohibition on mining natural resources is often expressed in terms of the violation of a woman's body.[4] Indeed the figuration of woman as being perhaps metonymically associated with Heimat and by implication with nature persists through almost all the twists and turns of the Heimat genre, even into Edgar Reitz's otherwise quite critical re-reading (Kahlenberg et al. 1985: 103).

In *Die Weiße Hölle vom Piz Palü* (Prisoners of the Mountain, dir. Arnold Fanck and G.W. Pabst 1929) the astonishing beauty of the mountain environment is coupled with terrible dangers: most of the mountaineers attempting to climb Piz Palü die. Dr Johannes Krafft loses his wife in a climbing accident immediately after being warned by his guide to curb his flippant attitude, which is apparently punished by the mountain. Two further disasters – the death of five students, and the accident which leads to Krafft, Maria (Riefenstahl) and her husband Heinz being trapped on a ledge – are also clearly due to over-ambitious and competitive behaviour. The students are keen to beat Krafft to the top, and so take a dangerous route blindly trusting that there will be no avalanche. Heinz (jealous after being forced to share the straw bed in the mountain hut with Krafft as well as his wife) insists on leading, to counter Maria's admiration of Krafft's skill. His lack of experience, however, causes the accident which eventually leads to Krafft's sacrificial death.

In contrast to the competitive young men, whose desire is simply conquest, whether of the mountain or Riefenstahl, Krafft has learned his lesson and has developed a particular affinity for this dangerous environment, so that he becomes more like the villagers, who are both knowledgeable about, and respectful towards, the mountain. Indeed, he goes beyond their respectful distance to become something of a 'spirit of the mountain' who wanders ceaselessly and rather obsessively on the slopes, tormented by the death of his wife. He finally finds peace by being left to die on the mountain rather than be rescued, supposedly sacrificing himself so that the young couple might survive, but mainly as a response to his own desire to become one with the mountain, frozen into its contours like his wife, whose body is also encapsulated in ice. Krafft's decision to remain on the mountain to die is his final journey towards a spiritual destiny of mystical unity with the mountain and by implication with his dead wife. Only in death it seems can there be a truly harmonious relationship between the male climber and his wild surroundings, equal to that of the wild woman in *Das blaue Licht*, who embodied the spirit of the mountain, and was so closely tied to the fate of her natural domain that she died when its mystical heart was ripped out.

Die weiße Hölle vom Piz Palü and *Das blaue Licht* rather ambiguously figure the wild mountain environment as a mystical locale infused with Romantic grandeur which separates it from the homely comforts of the tamed Heimat. Fanck, Trenker and Riefenstahl deliberately evoked the imagery and compositions of Romantic painters such as Caspar David Friedrich (Jacobs 1992: 32–35). This ultimately amounts to a rather problematic adulation of the mountain which certainly has its derivation in the early texts of such dubious Heimat prophets as Bartels, Lienhard and Langbehn. Mountain film aesthetics were also influential in the later development of a more explicit fascist aesthetic: Riefenstahl's infamous opening to *Triumph des Willens* (Triumph of the Will, 1935) showing Hitler apparently descending from the clouds owes much to shots designed by Fanck and also used by Riefenstahl. Siegfried Kracauer saw the mountain film as feeding into Nazi mythology in that it emphasised antirationalism, the lure of the elemental, and the value of sacrifice for a higher cause (Kracauer 1947: 110–12). However, in Fanck and Riefenstahl the alpine environment is not merely something to be conquered. The prohibition on mining implicit in *Das blaue Licht*[5] is reminiscent of conceptions of Gaia in nature cults and certain strands of today's mythically-coloured environmentalism (Sheldrake 1990). Similarly *Die weiße Hölle vom Piz Palü* emphasises the importance of respect for the mountain, as opposed to the complacent competitiveness of boy climbers who need to be taught a lesson. The films of Luis Trenker, however, are frequently rather more macho in mood, marginalising women characters and aggrandising the male climber, played by Trenker of course, who conquers the peaks and returns to the village to claim his Heimat maiden.[6] In *Der Berg ruft* (The Call of the Mountain, 1937) Trenker plays an Italian climber, Carrell, who plans to ascend the Matterhorn with Edward Whymper. Caught between various national interests which manifest themselves in an argument about whether they should begin their attempt from Switzerland or Italy, they climb separately. Whymper is first to the top, but is blamed for the deaths of some of his team members. Only Carrel can prove his innocence by finding the frayed rope that was allegedly cut. He risks his life to do this and the two friends finally climb the Matterhorn together. The central narrative of the film is the conquest of the mountain, first by Whymper, whose team nevertheless suffers fatal casualties on the return. This serves to render the second conquest by Carrel to save Whymper's reputation – solo and in foul conditions – even more heroic. The eventual victory of the hero over the mountain both saves the male bond between climbers and proves his worth to Felicitas, who has stuck by him when others doubted. In *Piz Palü*, the ambitious climb is never completed – rescue and a longed-for death on the mountain provide the denouement.

The key environmental problem looking retrospectively at these Alpine features is tourism – even as the beautiful images delight the urban public, they open up a fragile ecosystem to significant risk of degradation. The problem of mountain tourism as environmental threat, however, is not obvious in the 1920s and 1930s, with the possible exception of Riefenstahl's *Das blaue Licht*, where the artist, a well-meaning but naïve tourist, has initiated the destruction of the crystal chamber. Indeed in *Der verlorene Sohn* (The Prodigal Son, dir. Luis Trenker 1934), the criticism of the American woman who wants to be a mountaineer and nearly tempts Tonio from his true Heimat maiden centres on xenophobic anxiety about the incursion of the foreign into the Heimat, rather than a specific concern about the environment.

By the 1950s, the aesthetically innovative yet politically dubious mountain films, with their divisions between Heimat hearth and wild nature, had given way to ideologically conservative Heimat films, in which a rather sanitised view of rural spaces was sometimes opposed to the city, and sometimes to an idea of modernity. The lack of direct discussion of environmental dilemmas in 1950s Heimat films is hardly surprising, given that one of the key aims of the studios was to promote rural parts of Germany as holiday destinations (Rippey et al. 1996: 151–55). As the 1950s wore on, the dream of holidays and motor cars could be realised by more and more citizens, and Heimat films became serious shop windows for German resorts, alongside their other ideological functions of providing harmony and reconciliation for displaced populations, disrupted families, and alienated generations (Koch et al. 1989: 69–95). The potential for degradation of the very environment tourists come to see is rarely touched on in Heimat films. Cars for example, as an important object of desire in the consumer-oriented 1950s, must feature, though usually without any indication of potential damage to the beautiful surroundings. Two examples of 1950s films which do obliquely reflect environmental concern instead of just showing a rural Heimat untouched by modernity are *Dort oben, wo die Alpen glühen* (Up Where the Alps Are Glowing, dir. Otto Meyer 1956) and *Der Förster vom Silberwald* (The Gamekeeper of the Silver Forest, dir. Alfons Stummer 1954). In the former, the detrimental effects of tourism are discussed explicitly, with much discussion over the building of a new road which, it is argued, will bring many more visitors into the area and boost the local economy. There is a clear critique of those who stand to gain from this development. Tourism itself is shown in its best and worst aspects in two visitors, whose actions illustrate the dilemma facing rural communities. The likes of the young woman mountaineer, who respects the local environment and traditions, might cautiously be welcomed, but not her uncle, a comic figure, who seeks only to assert his superiority over local people and landscapes.

In *Der Förster vom Silberwald* there is an emphasis on the conservation of habitats by hunters and gamekeepers. This is opposed in the film to certain economic interests in the local community, where some would like to sell timber rights to the Silberwald. Though hunting has the macho overtones of conquest evident in Trenker's mountain conquests, and today its claims to conservation and the protection of wildlife have been challenged, the management of habitat by a gamekeeper is preferred in the film to the outside exploitation of the woodlands, which would threaten the destruction of a valuable habitat. The film's plot has Liesl torn between a Heimat life and the love of Gerold the gamekeeper, and a bohemian existence in Vienna living with an artist. The artist attempts to prove himself to her by engaging in country pursuits such as shooting. However, his lack of knowledge of Heimat ways finds him out, as in his ignorance he shoots a prize stag. Gerold generously covers for him, resigning from his post in the process. (His noble gesture does, however, win him the girl.) Again the thematic concern with conservation emerges from a suspicion of the outsider which is a staple of Heimat films. However, an irony of production is that *Der Förster vom Silberwald* was conceived originally as a promotional nature film to boost tourism in the area. It was only later given a plot as it was felt that it might then be even more effective (Seidl 1987: 82–3). After all, then, the meddling townies are wanted in the Heimat – or at least their much-needed money is, if the community is not to sell out to timber companies. The have-your-cake-and-eat-it approach of the 1950s films thus neatly if unintentionally encapsulates a key environmental dilemma: tourism threatens local environments but might save rural communities financially, and it means that beautiful habitats must be maintained so that there is something for tourists to admire. However, the destructive effects of mass tourism on these very habitats are only faintly alluded to, if at all.

Herzog and Reitz: The New Heimat Film

Following the angry rejection of the province in anti-Heimat films such as Peter Fleischmann's *Jagdszenen aus Niederbayern* (Hunting Scenes from Lower Bavaria, 1969) and Rainer Werner Fassbinder's *Katzelmacher* (1969), the 1970s saw a more differentiated approach as a political culture of environmentalism gathered momentum in the Federal Republic, fuelled both by the aftermath of the 1968 student revolt – which Colin Riordan identifies as feeding directly into the modern German Green Party (1997a: 32) – and by the growing interest in grass-roots democracy and local citizens' initiatives (Bahlinger et al. 1989: 146-47). New Heimat films emerged that did not utterly reject the province but were aesthetically and politically far from the reductive transmission of ideology that had been apparent in the 1950s films. Werner Herzog's films are politically ambiguous and suggestive of the kind of mystical attitude to nature that is evident in *Das blaue Licht* and *Die weiße Hölle vom Piz Palü*. They have aroused controversy for their sometimes bizarre experimental methods, such as the decision to film *Herz aus Glas* (Heart of Glass, 1976) with the entire cast bar Josef Bierbichler under hypnosis, and Herzog became notorious for exploitative behaviour in Peru during the making of *Fitzcarraldo* (1982). He went on in 1984 at the height of the German Green movement to make *Wo die grünen Ameisen träumen* (Where the Green Ants Dream), a film set in the Australian outback which directly thematises environmental concerns. However, even a relatively sympathetic critic like Thomas Elsaesser notes that Herzog's representation of native peoples is problematic (1986: 149).[7]

Herz aus Glas is a critical Heimat film, but not in the mode of Schlöndorff's *Der plötzliche Reichtum der armen Leute von Kombach* (The Sudden Wealth of the Poor People of Kombach, 1970), which uses authentic historical documents to narrate a story of exploitation. It is by comparison rather vague in its evocation of time (pre-modern) and place (Bavarian forests and language). In his evocative images of misty pasture, forests and the herding of cattle, Herzog recalls certain stock images of the Heimat film. However, landscape shots – wide horizons, swirling clouds and dramatic skies – are most often associated with the character of Hias, the shepherd-seer, and perform quite a different function to that in 1950s Heimat films, where natural surroundings are often rather anodyne and comforting, rather than unsettlingly mystical. The film concerns the aftermath of the death of a craftsman, the only person to know the formula for a highly-prized ruby glass. The owner of the glass factory desperately seeks written evidence of the formula, and also tries to reconstruct it himself. His obsessive quest leads him in his insanity to murder his servant girl, believing that her blood is the secret of the formula. Finally he burns down the glass factory, thus fulfilling one of Hias's prophecies. The apocalyptic vision of the clash of civilisation and nature is reminiscent of the environmental pessimism to be found in other critical Heimat texts, most notably those of Herbert Achternbusch, who wrote the script for *Herz aus Glas*.[8] But while Herzog might be the most serious candidate for a film-maker who reflects the widespread urgency about environmental issues evident in Germany in the 1970s and 1980s, his films are scarcely environmental campaign material; indeed while his interest in nature and landscape has inspired the admiration of American film-makers like Francis Ford Coppola, his attitude to the landscapes he so powerfully represents is not one of an earnest conservationist. He resisted strongly the suggestion that *Wo die grünen Ameisen träumen* should be read as just an environmental film, and he often suggests in his narratives and images that pure, wild nature is something sinister and dangerous. There is not really a sense of a fragile ecosystem that needs care and protection (Cheesman 1997: 292). His films do however open up the landscape to a grand vision and then thematise the relationship of 'culture' or 'civilisation', or even the individual, with the natural surroundings. Landscape dominates *Aguirre, der Zorn Gottes* (Aguirre, the Wrath of God, 1972): 'before there is character there is landscape', as Dana Benelli puts it (1986: 92). The film opens with an extreme long shot of the mountain slowly revealing a tiny chain of human beings descending to the river. As the conquistadores undertake their journey down the Amazon they are constantly at risk from the dangers of the mysterious and frightening jungle environment, be it dangerous river eddies or deadly fevers, even as they occasionally marvel at some natural wonder, like the butterfly that sits happily on a shoulder and the baby mammals that Aguirre shows to his daughter. Herzog does not stop at the earthly environment and its effects on humans – he also develops a cosmic perspective with the sky and specifically the sun suggested as a kind of father to the conquistadores: they are (mis)recognised as 'sons of the Sun' by the Indians who approach their boat, and in the final sequence where Aguirre is apparently the last person alive on the raft, a shot of the sun immediately follows his cry of 'Who is with me?' (Benelli 1986: 93).

Wo die grünen Ameisen träumen is a more apparently straightforward environmental narrative. The attempt by Aborigines to defend their sacred lands from white mining interests draws on the ancient and culturally widespread prohibition on mining the earth as a violation of an established natural order. But Herzog resisted the idea that his film could be read as simply a green fable:

> I don't see it as an environmental film, it's on a much deeper level: how people are dealing with this earth. It would be awful to see this film only as a film on ecology. It has a common borderline with that [but] it's also a film on a strange mythology, the green ants mythology. It's a movie, that's the first thing. (cit. Elsaesser 1986: 136)

Ecology of course has at its heart 'how people are dealing with this earth'. This relationship is central to Herzog's films, and so they do constitute an exploration of attitudes, images and myths that are part of an environmental awareness. However, they certainly resist categorisation as ecological films, and have little to do with the politics of environmental protest, even in the case of *Wo die grünen Ameisen träumen*. In a complex anthropological gesture the Aborigines are simultaneously the object of Herzog's western camera eye and autonomous subjects. As Elsaesser argues, there is an inversion of the ostensible view, so that even as we look at them we see civilisation, capital, and the geologist Lance Hackett from their perspective (1986: 145). In his figuration of the Aborigines Herzog implies a response to natural surroundings that goes beyond the 1970s and 1980s politics of environmental protection to depict the Aborigines, not just as 'protesters' hoping to guard the environment in the way of western greens, but as a part of the earth itself:

> Aborigines are the rocks you have to move away. They understand themselves as a part of the earth... That's why a man like Sam Woolagoocha said to me one day: 'They have ravaged the earth and don't they see they have ravaged my body?' That explains everything ... they are the rocks, they are the trees and you'd have to shoot them first, or blast them, before they would move. It has nothing to do with modern 'sit-in' techniques. (cit. Elsaesser 1986: 146–47)

So beyond the radicalism even of today's New Age environmental protesters, the radical identification of the Aborigines with the earth means that Herzog's film cannot after all be neatly coopted for green politics, for in this very identification lies the (for western liberals) uncomfortable representation of the Aborigines as somehow reified, lacking 'inner life' (Elsaesser 1986: 146), moved by Herzog in front of his camera just as the mining company seeks to move them out of the way (Elsaesser 1986: 149).

This uneasy and potentially dubious representation of human interaction or identification with nature is key to understanding in what sense Herzog is or is not relevant to debates about cultural constructions of the environment. In his films he pursues a vision of nature quite different to the idea of fragile ecosystem under threat. Nature is not only opposed to civilisation in his films, in some it threatens symbolically to destroy civilisation, as in *Aguirre, der Zorn Gottes* and *Herz aus Glas*, in a reverse of the green vision of apocalypse. In *Wo die grünen Ameisen träumen* as in other Herzog films, critique of western civilisation is palpable in the mythological reading of Nature, which opens up Herzog's bombastic landscapes to the criticism that he is reverting to the dubious irrationalism that Kracauer identified in the mountain films of the 1920s and 1930s. The anthropological identification with the Other of civilisation is similarly problematic. Herzog's use of powerfully incongruous symbols of technology in his documentary-style landscapes (mining machinery, the military plane in the desert) adds weight to the view that his films reveal an antirational visionary aesthetic,[9] presenting a natural world that can only be approached by civilisation's others, or by the megalomaniacs who sacrifice their sanity for some sort of mythological unity with nature.

Herzog, then, for all his documentary vision and ability to direct the natural landscape, does not make films which fit neatly into the green politics of the 1970s and 1980s. They do not present the environment as something to be conserved and guarded – the natural world is in his films something uncompromising and dangerous which involves terrible risks for civilisation. Though it is possible to make a case for his representation of other cultures (Elsaesser 1986, Benelli 1986), it is easy to see why he has offended western liberals and the western left so often with his anthropological approach. His apocalyptic visions, his dominant and dangerous landscapes, interwoven with myths that structure human approaches to the natural Other, nonetheless challenge the viewer to consider the clash between nature and civilisation in ways that are more radical than the 1970s and 1980s green rhetoric of environmental protection.

Herzog's overt suspicion of technology in *Wo die grünen Ameisen träumen* does have ambiguous undercurrents in the representation of the monstrous aeroplane in the desert which for all its incongruity has a status as object of desire.[10] Mistrust of technology and the implied critique of civilisation might be allied to a Romantic strand in green thought, a legacy which is contested in modern green politics. In the same year as Herzog's *Wo die grünen Ameisen träumen*, Edgar Reitz's ambitious family saga *Heimat* (1984) considered the relationship within Heimat between nature and technology, in a realist mode that manages to indicate some of the green dilemmas and tentatively suggest how they might be solved. From the very first episode of the saga, set in 1919, to its end in 1982, the impact of technology on the specific rural context is evident in the narrative as well as in the filmic images of the natural and built environment of the Hunsrück. In the 1950s Heimat, modern technology (with the notable exception of the car) appears to belong elsewhere, and tends not to encroach on nature shots that function as eye-candy for those stuck in cities undergoing reconstruction. In *Heimat* even landscape shots reveal the revolution in communications, in that telegraph wires and poles often divide the shot and the low hum of the wires can be heard. The saga shows how the rural community is connected with the rest of the nation by the late but then rather rapid arrival of communications technology. Radios, telephones, cars, and even motorways become part of the fabric of the Heimat as military ambition forces the pace of rural modernisation.

Heimat carefully reveals the Janus face of modern technology: the utopian potential of new technologies appears alongside a clear indication that the militarisation behind them will result in the betrayal of that potential. Thus when Ernst flies over the village to drop carnations for Anton's bride at her proxy wedding, the image will flicker for the viewer from the utopian potential and sheer beauty of flight to the violence of military bombing campaigns. In revealing the ambiguities of technological advance and exploring the interaction of modernity and the rural Heimat, Reitz is not simply engaging with the rural environment as threatened and fragile nature, but is looking particularly at Heimat as a site of human activity and network of social relations. The notion of environmental threat in the form of inappropriate developments is clearly present, indeed the rehabilitation of Heimat in the 1980s as a centre of identity worthy of recuperation is partly dependent on the growing interest post-1968 in ecology. However, Reitz also envisages a negotiation in rural spaces between conservation and the need for a local economy, often suggesting that technology's Janus face might be a potential saviour as well as destroyer of the environment.[11]

In Reitz's postwar sections, for example, the villain is not industry, but agriculture. The one former SS man in the village, Wilfried Wiegand, is the biggest landowner in the area, and his unseemly enthusiasm for agro-chemicals is set against Anton's clean optics factory. Anton deliberately places his factory in the Hunsrück environment because optics production requires clean air and argues bitterly with Wiegand, whose pesticide spraying forces Anton to stop production. The optics factory provides a considerable amount of skilled work for local people, in contrast to Wiegand, who merely pollutes the atmosphere and makes virtually no investment in the local economy. Anton's factory, which operates in harmony with the local environment, suggests not a simplistic ideal of preserving the Hunsrück as nature intended, but a more complex engagement with the local environment which takes into account the need for a sustainable local economy. Anton is no hero, environmental or other- wise – far from it, for his arrogance and intolerance draw him into conflict with the sympathetic Hermann – but he does hold out against the hostile bid of a multinational company, maintaining his focus on Research and Development and high-quality, small-scale production. His fragile success suggests, albeit rather tentatively, that rural areas need not simply be either agro-prairies or nature reserves, but that a careful expansion of a mixed rural economy could provide decent employment and still respect the environment.

In a further key contrast in attitudes to the rural Heimat, the avant-garde composer Hermann is set against his brother Ernst, the former Luftwaffe pilot. Both are rather alienated from the Heimat and experience their rural background as somewhat stifling, but Hermann responds by making sound recordings of the natural envi- ronment (birdsong) and creating a concrete, alienated Heimat music which expresses his debt to the natural and social site of iden- tity but in the same gesture transcends it. Hermann functions as a kind of alter ego for the director Reitz, who it is implied is also involved in expressing respectful indebtedness and yet desire for transcendence born of a critical and differentiated appraisal of Heimat. Ernst's response to his locality is cynical rather than criti- cal: he cons villagers out of their valuable antiques and literally dismantles the fabric of the Heimat, selling traditional rustic façades to theme bars in nearby cities. The heritage industry is certainly not an answer to the problems of local economies – Reitz displays a proper concern that the built as well as the natural environment should be respected, while satirising the modish (urban) desire for rustic chic. His critique is levelled not only at the absurdity of Düsseldorf theme bars, but also at a more serious example of heritage conservation, namely the wealthy Paul's bestowal of museum status on the family home, which is staged as pompous and faintly ludicrous.

The rural environment exists then in this more subtle Heimat film as a natural space under threat, a source of artistic inspiration, which even in the late twentieth century cannot be ignored, and finally as a place of social identity where the challenge is to find a livelihood for local people that might harmonise with the needs of the environment. Reitz weighs up critical Heimat art against heritage, and tries to envisage a green economy that could benefit the community as against earlier conceptions of an environment that had to be conquered, consumed or protected. His grappling, however tentative, with the notion of rural environment as a space to live and work is an interesting contemporary counterpoint to Herzog's exploration of apocalypse, mythology and cosmic mysticism.

In the 1990s, as green issues became more mainstream, environmental strands in films have become integrated into a range of different films which are not primarily films about ecological subject matter. At the same time, the mode of representing environmental concern has become rather more fragmented. Ecological concerns are evident in a number of films from the eco-disaster *Nach uns die Sintflut* (After Us the Deluge, dir. Siggi Rothemund 1991) to documentaries like *Die Wismut* (Wismut, dir. Volker Koepp 1993), emerging from the GDR tradition of documentary filmmaking, and as a caricatured *Szene* backdrop to the irritating comedy *Härtetest* (Trial by Fire, dir. Janek Rieke 1997). In a more serious mode, Hans-Christian Schmid's *23* (1998) also has the radical eco-protest scene as part of its context for examining the life of the young hacker Karl Koch. Even in recent work, however, one of the most important vehicles for examining the social context of interaction between people and sensitive environments remains the Heimat mode.

Tom Tykwer's 1997 film *Winterschläfer* (Winter Sleepers) alludes to the mountain film tradition and shows a traditional rural family failing to thrive after the disaster of losing a child in an accident. Tourism and particularly skiing are now fantastically more lucrative than farming. The farmhouse interior, reminiscent of Wilhelm Leibl's paintings, is contrasted with the luxury house plus indoor swimming pool of the ski resort manager. Here too the issue of how a local community can survive economically while still retaining a distinct character is an issue. Tourism and skiing bring work and money but have an uncomfortable impact on a small community. However, traditional rural industries do not seem to be realistic options for the four main characters in the film – a nurse, a translator, a ski instructor, the projectionist of a local cinema. The mountains are filmed with the same aesthetic care as in the earlier mountain films (missing from the vast majority of the 1950s Heimat films) and *Winterschläfer* echoes too the mystical longing for oneness with the mountains in death seen in *Die weiße Hölle vom Piz Palü*. Marco, the ski instructor whose complicated sex life brings nothing but misery, finally achieves an ecstatic moment of unity with the mountains as he falls to his death; the staging and editing of the fall suggest that Marco has achieved an everlasting moment of epiphany as shots of his accident are stretched out between shots of other characters living their lives over the next year or so.

While it is difficult to think of a specific body of films which deal explicitly with environmental thinking, ecological dilemmas and debates have often been expressed through the medium of the genre of the Heimat film, which has as one of its core concerns the interaction between human community and spatial location. This filmic tradition, which has existed since the mountain films of the Weimar Republic, was most closely associated with culturally conservative or even fascist politics until its appropriation by the left in a critical mode in the late 1960s and early 1970s. The dubious political tradition of the genre has meant that there is not always a particularly neat fit between post-1968 green politics and the critical Heimat films of Herzog, Reitz, Achternbusch, and latterly Tykwer, but in narratives, mise-en-scène and aesthetic style their films have drawn on some of the most diverse strands of green and proto-green thinking: the mythological prohibition on mining; the mystical desire for oneness with the mountains; Nature as Romantic ideal; a cosmic-scale vision of apocalypse; the powerful symbolism of dying forests; the dilemma of how a local community might interact harmoniously with a sensitive environment and yet still make a living. These films resist categorising as ecological or green films, but cultural and political environmentalism and proto-environmentalism are clearly an influential context, without which their diverse reflections on human interaction with nature remain difficult to understand. Conversely, their different visual conceptions of environment challenge spectators to think of the environment in ways that are complex and radical and go beyond the scope of much modern green campaigning.

Notes

1. An interesting source of information about documentary and other films dealing with environmental issues is the Ökomedia international film festival held annually in Freiburg im Breisgau. Non-German films outnumbered German films, however, in the 1999 competition, and most of the prizes were awarded to documentaries. See Internet <http://www.oekomedia-institut.de/FESTIVAL/index.html>.
2. Environmental debates in feature films in the 1970s and 1980s must also be seen alongside contemporary successes in documentary film-making, though there is not space to discuss these films in detail here. Such films were often broadcast on television, causing a much more palpable public impact than New German Cinema critical Heimat films. A notable documentary feature film in the eco-disaster mode is *Smog* (dir. Wolfgang Petersen 1973). I am grateful for Markus Kellermann for discussing these issues with me.

3. For examples of this tendency and a defence of Heimat societies based on their proto-green activism, see Rollins (1996) and Jefferies (1997).

4. See Carolyn Merchant 1983: 29–41. Hartmut Böhme also explores in detail the early modern characterisation of mining as the rape and matricide of Mother Earth. His 1988 book *Natur und Subjekt* is out of print but can be consulted on the Internet. See the chapter 'Geheime Macht im Schoß der Erde: Das Symbolfeld des Bergbaus zwischen Sozialgeschichte und Psychohistorie', <http:www.culture.hu-berlin.de/HB/Texte/natsub/geheim.html>, especially pages 5–8 and 26–7 of 57.

5. Eric Rentschler argues that the horror of nature defiled competes as a discourse in *Das blaue Licht* with the awareness of the commercial potential of the crystals for the villagers (1986: 171–173).

6. One of Trenker's most famous films, *Berge in Flammen* (Mountains in Flames, 1931), was allegedly adapted from a screenplay of Fanck's entitled *Die schwarze Katze* (The Black Cat). Based on the experiences of Fanck's cameraman Schneeberger in the First World War, it would have told the story of a woman skier and climber who risked her life to warn Austrian soldiers in a remote mountain base of an impending attack by the Italians. In Trenker's version the woman character is replaced by a male hero, played by Trenker himself (Riefenstahl 1992: 72–3).

7. For a hostile critique of Herzog as neocolonialist, see Davidson 1993; for a differentiated and perceptive defence see Koepnick 1993 and Cheesman 1997.

8. Achternbusch also works in the critical Heimat mode: in his 1984 film *Wanderkrebs* (The Spreading Cancer) he contemplates the environmental cause célèbre of the dying forest. He incorporates elements both of black humour and visual poetry, and in contrast to Edgar Reitz's tentative attempt in *Heimat* to reconcile the interests of environment and community, *Wanderkrebs* ends pessimistically with a lyrical sequence showing the suicide of the protagonist and his partner in a devastated forest (Pflaum and Prinzler 1992: 73–76).

9. Herzog is certainly indebted to the mountain film tradition – Rentschler suggests that *Herz aus Glas* perpetuates Riefenstahl's irrationalism in *Das blaue Licht*, which he associates with the elemental mythology of National Socialism (1986: 170–74).

10. See Leo Marx 1964 on the juxtaposition in literature of iconic symbols of industrial power such as the steam locomotive and an idealised natural landscape.

11. Luis Trenker has a perhaps surprisingly positive view of the use of technology in Alpine areas, as long as it is needed, and is integrated carefully; he criticises the tendency amongst some mountain climbers to demand that such regions be kept more or less as living museums and that local people be expected to manage without the modern conveniences the rest of us have come to rely on (Trenker 1961: 78–87, esp. 86).

CHAPTER 12

Children's Literature as a Medium of Environmental Education

Dagmar Lindenpütz

Nature and Childhood in Children's Literature

According to Philippe Ariès (French original 1960), childhood as an autonomous stage of life separate from adulthood is in Europe an invention of the seventeenth and eighteenth centuries. Prior to this children were considered to be 'amoral, not receptive to moral distinctions, "unformed"...' (von Hentig 1978: 10). They participated without restriction in the life of adults. With the transition from the 'big house' to the small (nuclear) family of the bourgeois age, however, there arose a new image of childhood: the child was no longer 'an object to pamper and play with, but innocent, corruptable, in need of protection and education, an object of grave responsibility' (von Hentig 1978: 10). The changed perception of childhood was reflected among other things in the literary market. In addition to numerous pedagogical treatises, the Age of Enlightenment also produced a wealth of texts intended to teach and entertain children. This period, where adult culture and children's culture began to separate, was therefore also the time which saw the genesis of children's literature as an autonomous literary genre – defined as the totality of texts which are 'expressly written, edited, compiled or published for children and young people' (Ewers 1990: 6f., Note 2).

For Rousseau the unspoiled nature of the child became a pedagogical leitmotif and the criterion for a sharp critique of civilisation. Since his revolutionary work *Emile* (written between 1758 and 1760), every age has taken up the challenge of redefining the position of the child in the area of conflict between nature and civilisation. The educational theorists of the Enlightenment – often also the enthusiastic authors of children's literature (e.g., Joachim Heinrich Campe, Christian Gotthilf Salzmann) – assumed a specific childlike nature which had to be refined in the process of education. The goal of the process was the mastery of internal and external nature. The Romantics saw things differently: seeking to counter the progressive demystification of the world by the sciences, they emphasised the sense children and poets have of the fantastic and mysterious (e.g., E.T.A. Hoffmann in his literary fairy tales *Nußknacker und Mausekönig* (Nutcracker and the Mouse King) and *Das fremde Kind* (The Strange Child), and stressed the close relationship between poetry and childish imagination. From this point of view, growing up is the history of the loss of creative potential. Finally, Biedermeier situated the experience of nature in a harmless idyll: 'Nature is loved because it is so far removed from the norms and conventions of civilisation, without making demands of its own. And in particular the life of children, childhood, is considered idyllic...' (Pech 1985: 12f.). Nature became the transfigured space of childish happiness, but nothing of it radiated out into the individual's own everyday life: 'Nature is reduced to spring, babbling brooks and young goats leaping boisterously, the emotional revolt against the dominance of reason and practicality becomes sentimentality.' (Pech 1985: 13). These Biedermeier stereotypes live on in contemporary children's literature.

Towards the end of the nineteenth century a new way of thinking about nature emerged, the Social Darwinist approach. It led to a dangerous equation of nature and society, culminating in the twentieth century in the racist ideology of National Socialism. The 'right of the strong' and the elimination of the weak, homeland and race, blood and soil – all these catchwords also found their way into children's literature (see Nassen 1993: 95–106). It is not surprising after the horrors which the National Socialists' politicisation of nature had given rise to that, in the postwar literature of the Federal Republic, nature and politics were again strictly separated. In West German children's literature of the 1950s nature is – just as in Biedermeier – primarily a space for children to play and be cosseted, it provides a safeguard against all too early integration in the adult world. It was precisely this remoteness from society in the natural idyll that triggered harsh criticism in the 1960s and 1970s among the authors of antiauthoritarian children's literature – the result being that the natural environment was still largely excluded as a subject in their books, and was replaced by an interest in the social environment. In view of the rapid increase in environmental destruction taking place at the same time, this produced a gap in perception, which ecological children's literature strives to fill. This new genre of text is characterised by the fact that it deals with current problems of the natural and social environment under the overall heading of ecological crisis. It reflects adults' fear of a global crisis and the imminent decline of humanity – combined with hope that the next generation will be more caring in their dealings with nature.

The Ecological Turnaround in Children's Literature in the Federal Republic

In the Federal Republic debate on nature conservation and the environment has taken on a specific tone related to Germany's past. Gudrun Pausewang, for example, warns against culpably producing a second disaster after that of National Socialism, against being blind a second time to a development which, if we keep our eyes open, we can predict and avert. She is afraid of again failing the generation of children who would inherit a devastated world, instead of teaching them harmonious coexistence – henceforth between man and nature. This adult fear of a new failure has helped create the situation in the Federal Republic, where great store is set by environmental education and apocalyptic perspectives have been the subject of intense discussion.

The debate on Germany's past is reflected, for example, in the subtitle of the book for young people *Die Wolke* (The Cloud) by Gudrun Pausewang (1987): 'Now we won't be able to say we didn't know'. The author is alluding to the widespread excuse used by Germans with regard to the Nazi period. Pausewang deliberately wants to provoke her young readers by giving them a horrifying depiction of the consequences of a fictitious 'maximum credible accident', or meltdown, in a German nuclear power station. Children – and parents (as the secondary targets of this work) – are to be made aware of the potential dangers of nuclear technology and to learn not to repress or trivialise them (see Dahrendorf 1991: 56–61).

In addition to the hazards emanating from nuclear power stations, nearly all the themes which feature in the ecological debates of adults can be found in West German children's literature, including the extinction of species and the destruction of animal and plant habitats, increasing road traffic, oil disasters, chemical accidents, ozone depletion and the greenhouse effect. There are certain adaptations and simplifications to adjust the problems to the awareness horizon of the target readers, but one basic principle applies: children are supposed to learn the truth in as comprehensive and unadulterated form as possible so that they can commit themselves to conserving their environment and take an informed part in public discourse on ecology.

Ecological works for children and young people have appeared in the Federal Republic since the 1970s. It is striking that the best known of these were initially translations (e.g., George, original US edition 1972, German translation 1974, Conti, original Italian edition 1978, German translation 1980). The illustrated books of the Swiss graphic artist Jörg Müller (1973, 1976) enjoy great esteem among the ecologically aware, as do the works he created jointly with the writer Jörg Steiner (1976, 1977, 1981, 1983, 1989). In the 1980s there was a sharp rise in the number of ecological books written by German children's writers. Environmental exhibitions and readings in libraries, environmental competitions such as those held by the *Börsenverein des Deutschen Buchhandels* (Association of Publishers and Booksellers of the Federal Republic of Germany) and individual towns, local environmental events, the publication of bibliographies of environmental literature by the newly created Environment Ministries on the Federal level and in the individual States – all these measures have raised the general interest in ecological matters and promoted ecological children's literature.

As a result of German reunification and the related economic problems, the themes of 'nature' and 'the environment' have receded somewhat, but they continue to be present in public discussion and children's literature. (See, for example: *dtv junior-Themenliste 1999* – a thematic listing of children's book titles supplied by the publisher as a selection aid for booksellers and teachers.) Some texts, such as *Die Wolke* by Gudrun Pausewang or the short story *Antennenaugust* (Antenna August, 1975) by the GDR author Kurt David, have now even attained the status of school classics. Ecological commitment is still a major aspect of books for young people in the nineties. (See Nöstlinger 1990, Boie 1990, 1993.) However, Nöstlinger and Boie signalise, like other authors from Western industrialised nations (e.g., Wheatley, original Australian edition 1987, German translation 1991, Gates, original English edition 1991, German translation 1995, George, original US editions 1992 and 1994, German translations 1993 and 1997), that they no longer hope for the fast and simple solutions which children's literature in particular liked to present in the first flush of enthusiasm for the Green Movement. There is now a general acceptance that environmental problems cannot be solved at a stroke, but that they will continue to accompany man in the new millennium.

Forms and Functions of Ecological Children's Literature in the Federal Republic

In the Federal Republic the public discussion on ecological matters took off in the seventies and very soon it spread to the domain of the educational sciences. In environmental pedagogy there arose a new educational field and the academics involved expressly called on the authors of children's literature to work within it: 'Broad scope is opening up for media educationalists, and especially for the authors of children's literature who consider it important to help achieve the objective of environmental education' (Eulefeld and Kapune 1979: 38). Children's literature was thus subsumed into environmental education and media didactics. This represents a – certainly not unproblematic – attempt at functionalisation, but one which Federal German children's book authors have initially been willing to go along with in numerous works of ecological education.

Since the 1980s, if not before, ecological topics have been present in nearly all genres of children's literature in the Federal Republic: numerous nonfiction books and series of books impart information on ecological interrelations. Similarly environmental plays (such as those of the Berlin Grips Theatre, see Schneider 1995, 154ff.) are aimed at promoting ecological knowledge and are also intended to encourage commitment among children. The beauty of nature and the problems of its conservation are also becoming a major focus of religious children's literature – under the heading of 'Save Creation'. Furthermore the rediscovery of the central status of nature in childhood is giving rise to a renaissance of nature poetry for children (see Kliewer 1995, Ewers 1995). Narrative works present a particularly extensive spectrum of interpretations with respect to the current ecological crisis. These range from the exemplary story (to propagate appropriate environmental behaviour and a new environmental morality) and adventure stories to novels of adolescence and science fiction (see Lindenpütz 1999).

Ecological children's literature is written in a quite specific context of social problems, for which the different authors offer their readers different causal explanations and solutions. So it would seem appropriate to create an initial overview of different kinds of text by developing a typology of these literary patterns of argumentation on the ecological crisis (see Lindenpütz 1999). The following classification is based on an analysis of about 200 works (including translations of books from other Western industrialised nations) containing ecological subject matter which have appeared in the Federal Republic since 1970. In drawing up the typology, no account is taken of those works (including many children's detective novels) which have quite clearly latched onto 'nature and the environment' merely because this is a fashionable topic, without contributing in any way to the recipients' ecological education. In the following classification it can be seen that all three patterns of argumentation described can be linked with certain strands of literary tradition, each of which treats nature in a specific way.

Works of Ecological Didacticism

An initial group contains those works aimed at educating the target readers on ecological matters in an entertaining way. It includes factual narratives, stories or plays in which exemplary behaviour in the domains of 'nature conservation and environmental protection' is presented, as well as socially critical environmental poems. The authors regard the widespread lack of scientific and/or social *knowledge* as the central cause of the ecological crisis, and it is precisely this deficiency that they wish to rectify with the help of their works. A historical model here is the exemplary story so popular with the children's authors of the Enlightenment, nourished as it is by the optimistic Socratic view that knowledge of truth automatically entails correct action.

Works to Provide an Ethical Foundation of Environmentally Friendly Behaviour

A second group of works is characterised by the fact that their authors propose a quite different approach to solving ecological problems. In their view it is not a question of imparting ecological knowledge by means of more education, but of a change in conscious values. In this they follow those theoreticians (such as Drewermann 1991: 115) who stress the primacy of the inner bond between man and the cosmos: it is not lack of knowledge that has led to the crisis and that holds man back. There is now definitely sufficient knowledge to find a remedy, but there is a lack of *emotional* willingness to conduct a radical rethinking. Minor reforms which do not involve change in the belief in growth and progress are not enough. What is called for is a fundamental change of values based on a holistic awareness, on a feeling for the unity of man and nature. The authors of children's literature who follow this line of argument are seeking to arouse in their readers a sense of the inner bond with nature – primarily by means of sensitive nature poems and animal stories, and by wilderness myths and adventure stories – and thus to create the emotional foundation for an *ecological ethic*. This approach is influenced by Rousseau's ideas and those of the Romantics concerning the children's original closeness to nature, by the traditional image of the 'noble savage' (which in Germany was moulded very much by Karl May) and by modern spiritual interpretations of nature. In translations of North American works the Anglo-American 'land ethic' also plays a significant role.

Ironic and Apocalyptic Works Which Reflect the Failure of all Previous Attempts to Bring About an Ecological Turnabout

Compared with the two previous approaches, whose practicioners still hope to help bring about a general social transformation with their books, the authors of the third group of works have largely become resigned. They are close to the proponents of *postmodern positions* (e.g., Lyotard, Baudrillard) when they highlight the fact that our view of the world today is a pure media product and also stress the aspects of growing dehumanisation and chaotisation in their children's books – these being either playfully ironic or deeply pessimistic.

Enlightenment optimism, holistic ecologism, postmodern irony and scepticism – these are then the three typical basic attitudes which West German readers encounter in ecological children's literature. The difficulty for them, in view of the plethora of media offerings, is primarily how to obtain an overview of the positions put forward there. The authors themselves also have to struggle with the wealth of material, of course. Their problem within the information society is to put over to their readership the complex ecological interrelations in a way suitable for children without simplifying them inappropriately.

Ecological Children's Literature in the German Democratic Republic

The different nature of the political systems and world views, together with the different conditions of production for literature in the Federal Republic and the German Democratic Republic, have also influenced children's literature of course. Whereas in the Western industrialised nations so much was reported of environmental hazards that there was a danger of blunting public awareness, in the GDR there was little reliable information on the subject of 'ecology' up to and into the 1980s, and there was absolutely no broad public discussion. Children's literature quite clearly played a pioneering role here by giving information on ecological abuses and possible ways out (see Hormann 1995: 108ff.). Since the authors of children's books, as opposed to those writing for adults, were not subject to quite such strict censorship (see Dolle-Weinkauff and Peltsch 1990: 398), they were able to use this latitude to educate the public, and even initiated an ecological discourse. In this way they assumed a high degree of social responsibility and also bore a great personal risk. They ran the risk of being depicted as denigrators of the nation and of not being allowed to publish at all. It is thus understandable that they proceeded very cautiously in their works, endeavouring to give sound and balanced depictions and preferring to encode their social criticism. At the risk of sounding cynical, it can be said that the pressure of censorship in the GDR produced very thoughtful and sophisticated works of a high literary standard in ecological children's literature. At the same time the market in the Federal Republic, especially in the 1980s, produced many trivial stories not worthy of acknowledgement.

The background situation of the GDR authors was therefore a very different one from that of their Western counterparts. They had difficulty in gaining access to environmental information and were themselves subject to censorship, because in their works they were not permitted to question belief in the Communist Party, the planned economy and socialist progress. Despite these obstacles, however, the first books dealing with the damaging consequences for nature of increasing industrialisation or road planning appeared as early as 1969 (Nowotny, *Der Riese im Paradies* [The Giant in Paradise]) and 1970 (Beseler, *Die Linde vor Priebes Haus* [The Lime Tree in front of Priebe's House]).

Within the children's literature of the GDR, as in that of the Federal Republic, it is possible to distinguish three central modes of argumentation on the ecological crisis: firstly ecological reform of the existing political system, secondly rejection of the prevailing norms in favour of a new harmony with nature, and thirdly premonitions of decay and decline.

The Continuity of Socialism by Means of Ecological Reform

An author in the GDR like Wolfgang Spillner, who criticised the blindness of the functionaries with respect to ecological interrelations, did not have to reject the whole political system. Rather the conservation of nature – as common property – meant for Spillner a consistent further development of socialism: 'Conservation of nature and scenic riches for all! That is the constitutional task!' (Spillner 1984:149). Just like the proponents of *ecological education* in the West, authors such as Spillner (1977, 1979, 1984), Beseler (1970) and Wolff (1979) placed their hopes in the reformability of the state. They relied on the power of ecological instruction and through their books they sought to help open the political leadership to ecological arguments.

Nature as a 'Counter-World' to Existing Society

A second group of authors, on the other hand, believed that nature conservation could only be pushed through by drawing a clear line of demarcation from the prevailing values and customs. Nature here was the complete other as compared to society in the GDR, which was suffocating in planning and constraint: scope for imagination to unfold (Wellm 1974, 1983), or a 'realm of freedom' as against the state-controlled 'realm of necessity' (see Lindenpütz 1999: 189 ff.). The preservation of nature as a 'counter-world' (see Großklaus and Oldemeyer 1983) is at the same time an important precondition for artistic creation, which also preserves and transforms man's dreams of nature (Pludra 1980). In the combination of nature and freedom, in the emphasis on the emotional value of an intact environment there are in turn parallels to Western positions as put forward rooted in holistic notions of nature.

Premonition of Decay and Decline

Social decay (Koch 1988) and the danger of global environmental destruction (Abraham 1985) were also subjects which moved GDR authors. In his youth novel *Augenoperation* (Eye Operation) Jurij Koch describes environmental destruction as a symptom of general social decline in the GDR. The burning *Affenstern* (Ape Star) in Abraham's satirical parable becomes a symbol of the apocalypse which threatens to destroy the earth if no way is found of stopping the steady destruction of nature. Once again the analogy with apocalyptic scenarios in Western children's literature is unmistakable.

For all the differences between the political context in the Federal Republic and the GDR, many factual ecological problems depicted in the children's literature of the two states were identical. These included the extinction of species and the destruction of biotopes, the pollution of nature by industry, construction measures and road traffic, the educational deficit of the general public and the lack of interest in ecological interrelationships among politicians, the primacy of economics over ecology. The fear of a final catastrophe could not be presented without disguise, but only in the unreal form of a story obviously born of the imagination –

Abraham (1985) demonstrates this more than clearly in his story *Der Affenstern* (The Ape Star). What was missing in the GDR was a debate on questions of nuclear and genetic engineering.

The works from the two German states share the triadic structure of the interpretation models for the ecological crisis they offer their readers. Either the authors see the main cause of the present environmental problems as lack of knowledge, and therefore plead for comprehensive education as the basis for the ecological reform of society. Or they stress the emotional significance of nature, and demand a radical conversion to a new holism (in the West) or the preservation of nature as a counterworld to the economic realm (in the East). Or they interpret the danger to the environment as a symptom of social decay and as the harbinger of global decline. Their resignation, their fear of, or ironic playing with, catastrophe place them close to postmodern positions.

Examples of Works and Publishers

In the 1980s the Federal German publishing house Erika Klopp Verlag began publishing a whole series of ecologically educational stories for children aged between about eight and twelve. Most of these works were brought out some time later in paperback by Deutscher Taschenbuch Verlag. Indeed it is to the great credit of this publisher that within its paperback series 'dtv junior' it offers a wide range of ecological literature for children and young people.

The educational stories published by Erika Klopp Verlag and Deutscher Taschenbuch Verlag deal with all those problems of species and biotope conservation which the authors and publishers hope will provoke a positive reaction from the children. The topics include, for example, the protection of beavers, storks or wild cats (Fischer-Nagel et al. 1986, 1989, 1992) or the conservation of valuable habitats such as so-called *Streuobstwiesen* (meadows with free-growing fruit trees) (Scherf 1991). Dust jacket texts, forewords and epilogues mark out the pedagogical framework for these works. The material ecological problems dealt with are explained to the target readers (and the adults as potential buyers of these books), and it is also stressed that the cases depicted are close to reality, that the authors themselves are knowledgeable in this field and that the child protagonists set an example by committing themselves to natural and environmental protection. Sometimes members of actual nature conservation organisations appear as figures in the plot. Occasionally the accompanying texts promote these organisations directly, e.g., the World Wildlife Fund or *Bund für Umwelt- und Naturschutz* in Germany, and even call for cash donations (see

Lindenpütz 1999: 77). By adopting something of a sledgehammer approach in environmental education, these works tend to just miss their target. The children acting as outstanding exemplary figures and their achievements are often very remote from the everyday life of the recipients, and so no transfer is possible. Should the recipients really attempt to imitate the protagonists, their disappointment is a foregone conclusion. If, on top of this, the animals shown as being under threat are also unacceptably trivialised and humanised (e.g., in Winsemius 1983), it can be assumed that the young readers will not develop an appropriate attitude towards them (see Lindenpütz 1999: 82).

The story *Antennenaugust* by the GDR author Kurt David is conceived of in a quite different way to most West German educational stories. From the child's naïve perspective the ignorance and irresponsibility of adults are unmasked in their dealings with nature. The story is about a boy who is entrusted with the rearing of a buzzard. He fails in the attempt to return the bird to the wild after it has become accustomed to the company of humans. It is left up to the reader himself to reconstruct the reasons for this failure, he is challenged to form his own judgement on the biological and social facts depicted. Like David, the West German woman author Gudrun Pausewang also promotes independent thought on the part of her recipients with her collection *Es ist doch alles grün* (But it's all green, 1991). Many of the texts in the collection give an exact depiction of positive examples for ecological action, like the educational stories mentioned above published by Erika Klopp Verlag and Deutscher Taschenbuch Verlag. However, in this case the author applies strategies of allusion and omission, using linguistic images and symbols to create free space for interpretation which readers have to fill out themselves with their own imagination. She thus provides them with the freedom of reception, instead of robbing them of their independent judgement by means of narrow didactic guidance.

The works published by the Swiss house of Sauerländer have encountered a very positive reaction in the Federal Republic. Without exaggeration, they can be described as representing the avant-garde of ecological children's literature in Europe. Sauerländer publish, for example (partly together with Aare Verlag), the author Jean Craighead George and they have therefore provided the German-speaking public access to the knowledgeable and exciting books of this American ecologist and writer of youth literature. In 1973 Sauerländer published the first comprehensive ecologically educational story in German with the translation of George's *Who Really Killed Cock Robin*, originally published in the United States in 1971. But George mainly became known with her book *Julie of the Wolves* (1972, translated into German in 1974). In it she depicts the odyssey of an Inuit girl through the Arctic tundra where she has got lost. Everything worth knowing about the ecology of this habitat and the ethology of its animal inhabitants is made clear from the perspective of the observing girl, who has to 'study' the behaviour of the wolves because she needs their help to survive. For this book the author was awarded the German Youth Book Prize in 1975, the highest award for children's literature in Germany. In addition to *Julie of the Wolves*, other exciting ecological tales of the wilderness by George (1985, 1988) have appeared in German translation.

German authors, on the other hand, have their difficulties with the wilderness narrative, not least because there has been no actual wilderness in densely populated Central Europe for centuries and – since there appears to be an absence of dominant nature – there is also no need to try and achieve a symbiotic relationship with nature. The depiction of nature reserves and nature 'paradises' by German authors is not infrequently linked with a happy holiday atmosphere, strongly reminiscent of the postcard idylls we see in tourist advertising.

One of the German-speaking authors who manages to give a convincing depiction of harmony between man and nature, such as we otherwise only find in American wilderness myths, is Walter Thorwartl (1990) with his story *Der Luchsfelsen* (The Lynx Rock). The mythical experience of oneness with nature, however, is restricted to a brief moment in a feverish dream. As she lies sick in bed, the girl Bernadette dreams of the lynx whom she wishes to save in his habitat, the rocks, and protect from the hunter. At the same moment the hunter actually misses the creature because he imagines he sees the girl standing next to it.

The most popular medium for depicting the harmony of man and the cosmos and the inner bond between all living things is still the nature poem in Germany. Josef Guggenmos, born in 1922 in the Allgäu, has earned considerable recognition for his children's poetry (and many literary awards). He works indefatigably at sensitising the young generation to nature, e.g., in the following poems:

Winter Forest
I walk through the snow
along the forest track. All that has
gone before me is the fox.

June Evening
Around me tiny glimmering
stars; glow-worms play
the cosmos at the forest's edge. (Gelberg, ed. 1993: 221, 247)

In the GDR it was mainly two authors who dealt with the fragile relationships children have with nature – constantly questioned as they are by busybodying and intrusive adults: Alfred Wellm with his sensitive picture books *Das Pferdemädchen* (The Horse Girl, 1974) and *Das Mädchen mit der Katze* (The Girl with the Cat, 1983), and Benno Pludra with his penetrating novel *Insel der Schwäne* (Island of Swans, 1980). Wellm has come under serious attack from literary critics for his nature-friendly attitude: 'Goethe's maxim: Let man be noble, generous and good! is difficult to apply to the relationship between humans and animals without restricting, even jeopardising man's habitat and existence.' (E. George 1987: 57). GDR literature of the 1980s still had the precept of 'socialist realism' hanging over it, i.e., the requirement to depict one's own society in terms of socialism and to participate in its further development in accordance with the guidelines set by the Communist Party. Nature and imagination as spaces of play and *refuge* in contrast to a social situation perceived as unsatisfactory (Wellm, Pludra), art as a medium for preserving human longing for *security in nature* (Pludra) – such notions, far removed from the planned economy and scientific-technological rationality, did not please the guardians of socialist progress.

Gudrun Pausewang was also attacked as an enemy of progress and a denigrator of the government for her criticism of West German nuclear policy in the novel *Die Wolke*. This book belongs to the third type of text in Western children's literature, the postmodern apocalyptic scenarios. For her novel the author won the German Youth Literature Prize of 1988, the only state prize awarded by the Federal Republic in the domain of youth literature. 'Since 1956 it has been awarded and financed by the Federal Youth Ministry of the day' (Voll 1991: 66). But this time the public was given the impression that the competent Federal Minister (Rita Süßmuth, CDU) only reluctantly presented the prize to Pausewang – although she had been chosen by an independent jury (see Voll 1991: 66–71). With her depiction of the unimaginable chaos that would break out after a 'maximum credible accident' in a nuclear power station in the Federal Republic of Germany and the critical picture she gives of the responsible politicians, the author becomes involved in the nitty-gritty of everyday politics. Pausewang's *Die Wolke* (see also her book on the consequences of an atomic war in Germany: *Die letzten Kinder von Schewenborn* [The Last Children of Schewenborn], 1983) and other critical youth novels on the subjects of 'global environmental destruction' (Heyne 1991, Schwarzer 1993), 'test-tube babies' (Kerner 1989) and 'genetic engineering/cloned children as a spare parts bank' (Rabisch 1992), show how far at least one part of ecological children's literature has left the clichés of a sheltered childhood and the Biedermeier natural idyll behind. Even in works for younger children, such as Michael Ende's environmental fairy story *Der satanarchäolügenialkohöllische Wunschpunsch* (The Satanarchaeolyingenialcohellish Wishing Punch, 1989) or the parable *Der Affenstern* (The Ape Star, 1985) by the GDR author Peter Abraham (both suitable for children from about ten years old), the deadly seriousness of the possible end of the world shimmers through the humorous and ironic play on the catastrophe.

If we consider the triadic typology developed above with a view to the chronology of the works, it can be seen that, with the emergence of the Green Movement in the 1970s, Enlightenment optimism initially predominated. It was hoped that energetic persuasion, among other things with the help of children's literature, would set the necessary ecological reforms in motion in the industrialised societies. This positive thinking was reflected in the educational stories for children. However, it was very soon found that, despite the growth in knowledge on the ecological crisis, the social transformation aimed at was very slow to get going. The number of voices calling for a radical rethink grew. It was pointed out that it was precisely rational, analytical thinking that was the reason for the present environmental problems and so the only way to overcome the crisis was to adopt a completely different, intuitive-holistic approach. This trend can also be seen in children's literature, in texts where the emotional experience of nature and a revival of mythical thinking are to the fore. Here a basis for ecological ethics was sought. Religious children's literature also took up this line of argument by seeking to arouse in its recipients joy in the beauty of creation and commitment to its preservation (see, for example, Bolliger and Capek 1981, Schindler and Heiduck-Huth 1982, Bolliger and Lemoine 1989, Pausewang 1990). For the proponents of the third mode of argumentation, however, both ways out – the creation of a new ecological rationality, and of a greater sensitivity to nature – are off-target. Like the authors of the second category, they have been increasingly evident since the 1980s. With black-as-night scepticism they depict a slide into the ultimate catastrophe (e.g., Pausewang 1983, 1987, 1991: 95-100, Schwarzer 1993). Even in the playfully ironic, somewhat more optimistic variants (e.g., Ende 1989), it remains unclear whether it would be possible to stop scientific-technological civilisation destroying itself.

Educational Versus Literary Value

Authors who place their works all too clearly at the service of a pedagogical idea are very quickly suspected in Germany of neglecting aesthetic quality. This debate on the primacy of aesthetics or pedagogy has accompanied children's literature since Heinrich Wolgast published *Über das Elend unserer Jugendliteratur* (On the Lamentable State of Our Youth Literature) in 1896, opposing the functionalisation of literature for non-literary purposes. What is the situation with regard to ecological children's literature? Does it succeed in reconciling ecology and aesthetics, or is it only possible to put over one at the expense of the other? The following attempt to answer this question is based on the propositions put forward by the present author in her doctoral thesis (Lindenpütz 1999).

First, it is undoubtedly legitimate to deal with ecological problems in children's literature, because they will be a central part of life in the future. Second, it is the task of literature to interpret human existence. Material problems of ecology and questions of environmental ethics can be highlighted with the help of exemplary stories, and social structures can be illuminated using fables and parables. The complex structures of reality are simplified here, but at the same time they are preserved in the literary image or parable. I suggest that children's literature makes a major contribution to ecological education by helping readers understand the problems of the world they live in. At the same time, it encourages independent thought by, for example, presenting the same event from different perspectives or by challenging readers with allusions, omissions or an open conclusion to fill out the gaps themselves. Finally, it is the inherent function of literature to stimulate the imagination, design utopias, or even fear-inducing dystopias to break through firmly established ways of thinking and habits in existing societies, and thus to precipitate a change for the better. The educational claim of ecological children's literature is based on the persuasive force of its images and a tight interlinking between the individual text elements to project the intended message. One major quality criterion is the reflected application of literary strategies which render the present social problems transparent to the recipients without falsifying them. It is precisely this that the authors of ecological children's literature succeed in doing in their best works (see George, US original 1972, 1983, 1987, Müller 1973, 1976, David 1975, Conti, Italian original 1978, Pludra 1980, Müller and Steiner 1981, 1989, Scholes and Buchholz, Australian original 1985, Abraham 1985, Ende 1989, Nowotny 1990, Hesse, US original 1994).

NOTES ON CONTRIBUTORS

Amery, Carl, nom de plume of the novelist, essayist, cultural critic and environmental campaigner Christian Mayer. Sometime member of *Gruppe 47*, President of the German Writers' Union and President of German PEN. Author of novels including *Der Untergang der Stadt Passau* (1975) and *Die Wallfahrer* (1986), and essays including *Die Kapitulation oder Deutscher Katholizismus heute* (1963), *Das Ende der Vorsehung* (1972), *Natur als Politik* (1976), *Bileams Esel* (1991) and *Botschaft des Jahrtausends* (1994). Has also written plays and film scripts. Lives in Munich.

Baukloh, Anja, political scientist, researcher at the Humboldt-University of Berlin. Principal fields of interest: political protest, civil rights and comparative policy analysis. Currently writing a doctoral thesis on the East German transition process in the 1990s. Recent publications include 'Truth and Secrecy – Ten Years into the Transition' (with Katy Crossley-Frolick and Matt Murphy), 'Nie wieder Faschismus! Antinationalsozialistische Proteste in der Bundesrepublik der 50er Jahre im Spiegel ausgewählter Tageszeitungen', and 'Portugal. Vom autoritären Korporatismus zum demokratischen Pluralismus' (all 2001).

Blühdorn, Ingolfur, Senior Lecturer in European Politics at the University of Bath, in the Department of European Studies. His research interests focus on ecological politics, particularly eco-political theory, social movements and the social theory of late modernity. Principal publications include *The Green Agenda. Environmental Politics and Policy in Germany* (co-edited with Thomas Scharf and Frank Krause, 1995), *Post-Ecologist Politics. Social Theory and the Abdication of the Ecologist Paradigm* (2000) and *Simulative Politics. Strategies for a Denucleated Modernity* (forthcoming).

Goodbody, Axel, Reader in German Studies, University of Bath. His research is principally concerned with twentieth-century German literature (nature poetry, Exile and GDR Studies) and film. Publications include: *Natursprache. Ein dichtungstheoretisches Konzept der Romantik und seine Wiederaufnahme in der modernen Naturlyrik* (1984), *Geist und Macht. Writers and the State in the GDR* (co-edited with Dennis Tate, 1992), *Umwelt-Lesebuch* (edited, 1997) and *Literatur und Ökologie* (edited, 1998). Currently writing a volume on *Twentieth-Century German Literature in Ecocritical Perspective.*

Hoffmann, Jürgen, Consultant and Head of Section for Universities in Southwest and Northeast Germany at the Konrad Adenauer Foundation, Sankt Augustin. Publications include *Die doppelte Vereinigung. Vorgeschichte, Verlauf und Auswirkungen des Zusam-menschlusses von Grünen und Bündnis 90* (1998) and 'Werden die Grünen überleben? Probleme einer Oppositionsbewegung an der Macht' (forthcoming).

Hope, Jacquie, Senior Lecturer in German at the University of Plymouth. Studied Modern Languages at the University of Oxford, and held a Junior Fellowship at Queen's College, Oxford, gaining her D.Phil. with *Green Trends in East Germany* (1992), a study of critiques of modern industrial society in GDR literature. Her research continues to focus on the literature of the former GDR.

Krönig, Jürgen, journalist, broadcaster, author and photographer. Has been the United Kingdom and Ireland correspondent for the German weekly *Die Zeit* since 1989, and works as a freelance author for TV and Radio as well as writing for other German, Swiss and British publications. Has written extensively about globalisation, the information technology revolution, media conglomerates and environmental issues. Recent publications include *The Secret Face of Nature* (2001).

Lindenpütz, Dagmar, secondary school teacher and member of the *Arbeitsgemeinschaft Kinder- und Jugendliteraturforschung.* Publications include *Das Kinderbuch als Medium ökologischer Bildung* (1999).

Palfreyman, Rachel, Lecturer at the University of Nottingham. Completed her PhD at the University of Manchester (*Edgar Reitz's Heimat: Histories, Traditions, Fictions*, 2000). Co-editor with Elizabeth Boa of *Heimat - A German Dream* (2000). Currently writing on the film director Tom Tykwer.

Rohkrämer, Thomas, Senior Lecturer in Modern European History at the University of Lancaster. His research is focused on German history in the nineteenth and twentieth centuries, in particular military and environmental history in a cultural perspective. Publications include: *Der Militarismus der 'kleinen Leute'. Die Kriegervereine im deutschen Kaiserreich* (1990) and *Eine andere Moderne?* (1999).

Roose, Jochen, researcher at the University of Leipzig, Germany. His work is concerned with environmental movements, Europeanisation, methods of social research and the sociology of organisations. Publications include 'The German nvironmental Movement at a Crossroads' (1999) and 'Neither Decline nor Sclerosis. The Organisational Structure of the German Environmental Movement' (2001 – both together with Dieter Rucht). His doctoral thesis on the Europeanisation of the German and British environmental movements is due to be published shortly.

SELECT BIBLIOGRAPHY

Abraham, Peter 1985 *Der Affenstern*. Berlin (East), Der Kinderbuchverlag

Adorno, Theodor Wiesengrund 1970 *Ästhetische Theorie*. Frankfurt am Main, Suhrkamp

Amann, Frank, Gabel, Ben and Keiper, Jürgen, ed. 1992 *Revisited: Der Fall Dr Fanck. Die Entdeckung der Natur im deutschen Bergfilm (Film und Kritik 1)*. Basel, Stroemfeld/ Roter Stern

Amery, Carl 1954 *Der Wettbewerb*. Munich, Nymphenburger Verlagshandlung

Amery, Carl 1958 *Die Große Deutsche Tour. Heiterer Roman aus den fünfziger Jahren*. Munich, Nymphenburger Verlagshandlung

Amery, Carl 1963 *Die Kapitulation oder deutscher Katholizismus heute*. Nachwort von Heinrich Böll. Reinbek, Rowohlt, 1963

Amery, Carl 1967 *Capitulation: An Analysis of Contemporary Catholicism*. Translated by Edward Quinn. With a foreword by Professor JM Cameron and an epilogue by Heinrich Böll. London and New York, Sheed and Ward

Amery, Carl 1972 *Das Ende der Vorsehung. Die gnadenlosen Folgen des Christentums*. Reinbek, Rowohlt

Amery, Carl 1974 *Das Königsprojekt*. Munich, Piper

Amery, Carl 1975 *Der Untergang der Stadt Passau. Science Fiction-Roman*. Munich, Heyne

Amery, Carl 1976 *Natur als Politik. Die ökologische Chance des Menschen*. Reinbek, Rowohlt

Amery, Carl 1979 *An den Feuern der Leyermark*. Munich, Nymphenburger Verlagshandlung

Amery, Carl 1985a *Die ökologische Chance*. Mit einem 'Nachwort 1985' (Reprint of *Das Ende der Vorsehung* and *Natur als Politik*). Munich, Süddeutscher Verlag

Amery, Carl 1985b *Die starke Position oder Ganz normale MAMUS. Acht Satiren*. Munich, Süddeutscher Verlag

Amery, Carl 1986 *Die Wallfahrer*. Munich, Süddeutscher Verlag

Amery, Carl 1987 "Wehe, wenn die Pilger los sind…" Interview with Bartholomäus Grill and Eduard Kopp', *Deutsches Allgemeines Sonntagsblatt* 3 (18 January), 16

Amery, Carl 1990 *Das Geheimnis der Krypta*. Munich and Leipzig, List

Amery, Carl 1991 *Bileams Esel. Konservative Aufsätze. Mit einem Vorwort von Walter Jens*. Munich, List

Amery, Carl 1994 *Die Botschaft des Jahrtausends. Von Leben, Tod und Würde*. Munich and Leipzig, List

Amery, Carl 1996 'Der Marsch in den Kompromiß', *Süddeutsche Zeitung*, Feuilleton-Beilage, 22/23 June

Amery, Carl 1997 'Ptolemäer und Plattweltler. Zur Kontroverse um die "Deklaration der Menschenpflichten"', *Die Zeit* 47 (14 November) 1997, 6. Reprinted in Günter Altner et al., ed. *Jahrbuch Ökologie 1999*. Munich, C.H. Beck, 1998, 22–6

Amery, Carl 1998 *Hitler als Vorläufer. Auschwitz – der Beginn des 21. Jahrhunderts?* Munich, Luchterhand

Amery, Carl 1999 'Zur Sonne, zur Freiheit. Hermann Scheer, neuer Träger des alternativen Nobelpreises, kämpft für die Solarenergie', *Die Zeit* 42 (14 October), 44

Amery, Carl, Mayer-Tasch, P.C. and Meyer-Abich, Klaus, ed. 1978 *Energiepolitik ohne Basis. Vom bürgerlichen Ungehorsam zu einer neuen Energiepolitik*. Frankfurt am Main, Fischer

Anders, Günther 1956 *Die Antiquiertheit des Menschen*. Munich, Beck

Ariès, Philippe 1978 *Geschichte der Kindheit*. Munich, dtv (French orig. 1960)

Bahlinger, Dieter, Hellmuth, Thomas and Reister, Tobias 1989 'Die achtziger Jahre: Nostalgie oder Neuanfang?' in Kaschuba et al., ed. 1989, 131–48

Bahro, Rudolf et al. 1966 'In diesem besseren Land', *Forum* 8, 19–23

Bate, Jonathan 1991 *Romantic Ecology. Wordsworth and the Environmental Tradition.* London and New York, Routledge

Bate, Jonathan 2000 *The Song of the Earth.* London, Picador

Bateson, Gregory 1979 *Mind and Nature. A Necessary Unity.* New York, Dutton

Batt, Helge-Lothar 1996 *Die Grundgesetzreform nach der deutschen Einheit. Akteure, politischer Prozeß und Ergebnisse.* Opladen, Leske und Budrich

Becher, Johannes R. 1959 *Ein Staat wie unser Staat.* Berlin, Aufbau

Bechtold-Comforty, Beate, Bedek, Luis and Marquardt, Tanja 1989 'Zwanziger Jahre und Nationalsozialismus: Vom Bergfilm zum Bauernmythos' in Kaschuba et al., ed. 1989, 33–67

Beck, Ulrich 1986 *Risikogesellschaft: auf dem Weg in eine andere Moderne.* Frankfurt, Suhrkamp

Beck, Ulrich 1993 *Die Erfindung des Politischen.* Frankfurt, Suhrkamp

Beck, Ulrich, ed. 1997 *Kinder der Freiheit.* Frankfurt, Suhrkamp

Beck, Ulrich, ed. 2000 *Die Zukunft von Arbeit und Demokratie.* Frankfurt, Suhrkamp

Beckerman, Wilfred 1995 *Small is Stupid. Blowing the Whistle on the Greens.* London, Duckworth

Benelli, Dana 1986 'The Cosmos and its Discontents' in *The Films of Werner Herzog. Between Mirage and History*, ed. Timothy Corrigan. New York and London, Methuen, 89–103

Beseler, Horst 1970 *Die Linde vor Priebes Haus.* Berlin (East), Der Kinderbuchverlag

Blank, Bettina 1998 'Der Protest gegen CASTOR–Transporte' in *Jahrbuch Extremismus und Demokratie*, ed. Uwe Backes and Eckhard Jesse. Vol. 10, 199–213

Bloch, Ernst 1976 *Das Prinzip Hoffnung* (First ed. 1959). Frankfurt am Main, Suhrkamp

Blühdorn, Ingolfur 1995 'Campaigning for Nature: Environmental Pressure Groups in Germany and Generational Change in the Ecology Movement' in *The Green Agenda. Environmental Politics and Policy in Germany*, ed. Blühdorn, Ingolfur, Krause, Frank and Scharf, Thomas. Keele, Keele University Press, 167–220

Blühdorn, Ingolfur 2000a *Post-Ecologist Politics. Social Theory and the Abdication of the Ecologist Paradigm.* London and New York, Routledge

Blühdorn, Ingolfur 2000b 'Ecological Modernisation and Post-Ecologist Politics' in *Environment and Global Modernity*, ed. Buttel, F., Mol, A. and Spaargaren, G. London, Thousand Oaks and New Delhi, Sage

Blühdorn, Ingolfur 2000c *Myths of Empowerment and Ecologisation. On the re-materialisation of post-materialist politics.* Paper presented at the ECPR Joint Workshops, Copenhagen, 14–19 April, 2000

Blühdorn, Ingolfur 2001 'Reflexivity and Self-referentiality: On the Normative Foundations of Ecological Communication', in *Language – Meaning – Social Construction. Interdisciplinary Studies*, ed. Grant, Colin and McLaughlin, Donal. Amsterdam and New York, Rodopi, 181–201

Bodenstein, Gerhard, Elbers, Helmut, Spiller, Achim and Zühlsdorf, Anke 1998 *Umweltschützer als Zielgruppe des ökologischen Innovationsmarketing. Ergebnisse einer Befragung von BUND-Mitgliedern.* Duisburg, Gerhard Mercator Universität Gesamthochschule

Böhme, Gernot 1989 *Für eine ökologische Naturästhetik.* Frankfurt am Main, Suhrkamp

Böhme, Hartmut 1988 *Natur und Subjekt.* Frankfurt am Main, Suhrkamp

Bohnke, Ben-Alexander 1997 *Abschied von der Natur. Die Zukunft des Lebens ist Technik.* Düsseldorf, Metropolitan

Boie, Kirsten 1990 *Das Ausgleichskind.* Hamburg, Oetinger

Boie, Kirsten 1993 *Jeder Tag ein Happening.* Hamburg, Oetinger

Bolliger, Max and Lemoine, Georges 1989 *Das Buch der Schöpfung.* Freiburg im Breisgau, Herder (French orig. 1987)

Bolliger, Max and Capek, Jindra 1981 *Am Anfang...* Zürich, bohem press

Bowie, Andrew 1998 'Critiques of Culture' in *The Cambridge Companion to Modern German*

Culture. Cambridge, Cambridge University Press, 132–152

Bramwell, Anna 1989 *Ecology in the 20th Century. A History.* New Haven and London, Yale University Press

Bramwell, Anna 1994 *The Fading of the Greens. The Decline of Environmental Politics in the West.* New Haven and London, Yale University Press

Brand, Karl-Werner 1999a 'Transformation der Ökologiebewegung' in: Klein et al., ed. 1999, 237–256

Brand, Karl-Werner 1999b 'Dialectics of Institutionalisation: The Transformation of the Environmental Movement in Germany' in Rootes 1999b, 35–58

Brand, Karl-Werner, Büsser, Detlef and Rucht, Dieter 1986 *Aufbruch in eine andere Gesellschaft. Neue soziale Bewegungen in der Bundesrepublik.* 2nd edition, Frankfurt am Main and New York, Campus

Bratton, Susan Power 1999 'Luc Ferry's Critique of Deep Ecology, Nazi Nature Protection Laws, and Environmental Anti-Semitism', *Ethics and the Environment* 4, 1, 3–22

Braun, Volker 1979 *Gedichte.* Leipzig, Reclam

Braun, Volker 1990 *Bodenloser Satz.* Frankfurt am Main, Suhrkamp

Brezan, Jurij, Peltsch, Steffen, Nowotny, Joachim et al. 1989 'Um Welt', *Neue deutsche Literatur* 11, 37, 5–150

Brüggemeier, Franz-Josef 1998 *Tschernobyl, 26. April 1986. Die ökologische Herausforderung.* Munich, dtv

Bundesministerium für Umwelt, Naturschutz und Reaktorsicherheit, ed. 1996 *Umweltbewußtsein in Deutschland 1996. Ergebnisse einer repräsentativen Bevölkerungsumfrage.* Bonn

Bundesministerium für Umwelt, Naturschutz und Reaktorsicherheit, ed. 2000 *Umweltbewußtsein in Deutschland 2000. Ergebnisse einer repräsentativen Bevölkerungsumfrage.* Berlin

Bürklin, Wilhelm and Dalton, Russell J. 1994 'Das Ergrauen der Grünen' in *Wahlen und Wähler. Analysen aus Anlaß der Bundestagswahl 1990*, ed. Max Kaase and Hans-Dieter Klingemann. Opladen, Westdeutscher Verlag, 264–302

Callenbach, Ernest 1975 *Ecotopia. A novel about ecology, people and politics in 1999.* Berkeley, California, Banyan Tree Books

Carson, Rachel 1962 *Silent Spring.* Cambridge, Mass., Riverside

Carson, Rachel 1990 *Der stumme Frühling.* Munich, Beck

Carter, Neil 1997 'Prospects: The Parties and the Environment in the UK' in Jacobs 1997, 192–205

Cheesman, Tom 1997 'Apocalypse Nein Danke: The Fall of Werner Herzog' in Riordan 1997a, 285–306

Chesterton, Gilbert Keith 1998 *Ketzer. Eine Verteidigung der Orthodoxie gegen ihre Verächter.* Frankfurt am Main, Eichborn

Christmann, Gabriele B. 1997 *Ökologische Moral. Zur kommunikativen Konstruktion und Rekonstruktion umweltschützerischer Moralvorstellungen.* Wiesbaden, Deutscher Universitätsverlag

Cibulka, Hanns 1982 *Swantow: Die Aufzeichnungen des Andreas Flemming.* Halle and Leipzig, Mitteldeutscher Verlag

Cibulka, Hanns 1985 *Seedorn: Tagebucherzählung.* Halle and Leipzig, Mitteldeutscher Verlag

Cibulka, Hanns 1988 *Wegscheide: Tagebucherzählung.* Halle and Leipzig, Mitteldeutscher Verlag

Commoner, Barry 1971 *The Closing Circle: Nature, Man and Technology.* New York, Knopf

Conti, Laura 1980 *Sara und Marco. Eine Geschichte aus Seveso.* Munich, Weismann (Ital. orig. 1978)

Cornelsen, Dirk 1991 *Anwälte der Natur. Umweltschutzverbände in Deutschland.* München, C.H. Beck

Cox, Harvey Gallagher 1965 *The Secular City. Secularization and Urbanization in Theological Perspective.* New York, Macmillan

Czechowski, Heinz 1981 *Was mich betrifft.* Halle and Leipzig, Mitteldeutscher Verlag

Dahl, Jürgen, Himmelheber, Max, Kragh, Gert, Lohmann, Michael and Schwabe, Gerhard Helmut 1975 'Bussauer Manifest zur umweltpolitischen Situation', *Scheidewege* 4, 469–486

Dahrendorf, Malte 1991 'Wider das Verdrängen und Verharmlosen. Gudrun Pausewangs

Atomkatastrophenbücher' in Runge 1991, 56–61

Dahrendorf, Malte, ed. 1995 *Kinder- und Jugendliteratur. Material.* Berlin, Volk und Wissen

David, Kurt 1975 *Antennenaugust.* Berlin (East), Der Kinderbuchverlag

Davidson, John E. 1993 'As Others Put Plays upon the Stage: *Aguirre*, Neocolonialism, and the New German Cinema', *New German Critique* 60, 101–130

de Bruyn, Günter 1992 *Zwischenbilanz. Eine Jugend in Berlin.* Frankfurt am Main, Fischer

Deubel, Werner 1934 'Auswirkungen des biozentrischen Weltbildes', *Süddeutsche Monatshefte* 31/4, 220–231

Diekmann, Andreas and Jaeger, Carlo C., ed. 1996 *Umweltsoziologie.* Special issue of *Kölner Zeitschrift für Soziologie und Sozialpsychologie*, 36. Opladen, Westdeutscher Verlag

Dierkes, Meinolf and Fietkau, Hans-Joachim 1988 *Umweltbewußtsein – Umweltverhalten.* Karlsruhe, Kohlhammer

Ditfurth, Jutta 1991 *Lebe wild und gefährlich. Radikalökologische Perspektiven.* Cologne, Kiepenheuer und Witsch

Dobson, Andrew 1995 *Green Political Thought.* 2nd edition, London, Routledge

Doherty, Brian 1999 'Paving the Way: The Rise of Direct Action against Road-Building and the Changing Character of British Environmentalism', *Political Studies* 47, 2, 275–91

Dolle-Weinkauff, Bernd and Peltsch, Steffen 1990 'Kinder- und Jugendliteratur der DDR' in Wild 1990, 372–401

Dominick, Raymond H. 1992 *The Environmental Movement in Germany. Prophets and Pioneers, 1871–1971.* Bloomington, Indiana and London, Indiana University Press

Dörfler, Marianne and Ernst 1986 *Zurück zur Natur.* Leipzig, Jena and Berlin, Urania

Downs, Anthony 1972 'Up and Down with Ecology: The Issue-Attention Cycle', *Public Interest* 28, 38–50

Drewermann, Eugen 1981 *Der tödliche Fortschritt: von der Zerstörung der Erde und des Menschen im Erbe des Christentums.* Regensburg, Pustet (2nd edition Freiburg im Breisgau, Herder, 1991)

Dunlap, Riley E. and Mertig, Angela G. 1996 'Weltweites Umweltbewußtsein. Eine Herausforderung für die sozialwissenschaftliche Theorie' in Diekmann and Jaeger 1996, 193–218

Ehmke, Wolfgang 1998 'Transformationen der Ökologiebewegung. Versuch einer Ökobilanz', *Forschungsjournal Neue Soziale Bewegungen* 11, 1, 142–153

Ehrlich, Paul R. 1968 *The Population Bomb.* New York, Ballantine Books

Elsaesser, Thomas 1986 'An Anthropologist's Eye: *Where the Green Ants Dream*' in *The Films of Werner Herzog. Between Mirage and History*, ed. Timothy Corrigan. New York and London, Methuen, 133–56

Emmerich, Wolfgang 1996 *Kleine Literaturgeschichte der DDR.* Leipzig, Mitteldeutscher Verlag

Ende, Michael 1989 *Der satanarchäolügenialkohöllische Wunschpunsch.* Stuttgart, Thienemann

Eppler, Erhard 1975 *Ende oder Wende. Von der Machbarkeit des Notwendigen.* Stuttgart, Kohlhammer

Eulefeld, Günter and Kapune, Thorsten, ed. 1979 *Empfehlungen und Arbeitsdokumente zur Umwelterziehung.* Kiel, IPN (=IPN–Arbeitsberichte, Vol. 36)

Ewers, Hans-Heino, ed. 1990 *Kinder- und Jugendliteratur der Aufklärung.* Stuttgart, Reclam

Ewers, Hans-Heino 1995 'Kinderlyrik als Naturlyrik. Vom romantischen Kindergedicht zur westdeutschen Kinderlyrik der Nachkriegszeit' in Nassen 1995, 177–197

Ferry, Luc 1995 *The New Ecological Order.* Chicago & London, University of Chicago Press

Findeis, Hagen, Pollack, Detlef and Schilling, Manuel 1994 *Die Entzauberung des Politischen. Was ist aus den politisch alternativen Gruppen der DDR geworden?* Leipzig and Berlin, Evangelische Verlagsanstalt

Fischer, Arthur, Fritzsche, Yvonne, Fuchs-Heinritz ,Werner and Münchmeier, Richard 2000 *Jugend 2000. 13. Shell Jugendstudie.* Opladen, Leske und Budrich

Fischer, Joschka 1984a 'Für einen grünen Radikalreformismus' in *Von grüner Kraft und Herrlichkeit.* Reinbek, Rowohlt, 119–135

Fischer, Joschka 1984b 'Identität in Gefahr' in *Grüne Politik. Der Stand einer Auseinandersetzung*, ed. Kluge, Thomas. Frankfurt am Main, 20–35

Fischer, Joschka 1989 *Der Umbau der Industriegesellschaft. Plädoyer wider die herrschende Umweltlüge.* Frankfurt, Eichborn

Fischer-Nagel, Andreas and Schmitt, Christel 1986 *Eine Biberburg im Auwald.* Berlin, Klopp (Munich, Deutscher Taschenbuch Verlag, 1989 [=dtv junior, Vol. 70161])

Fischer-Nagel, Heiderose and Andreas 1989 *Die Störche kommen!* Berlin, Klopp (Munich, Deutscher Taschenbuch Verlag, 1992 [=dtv junior,Vol. 70252])

Fischer-Nagel, Heiderose and Andreas 1992 *Wildkatzen in unseren Wäldern.* Berlin, Klopp

Frankland, E. Gene and Schoonmaker, Donald 1992 *Between Protest and Power. The Green Party in Germany* Boulder, San Fransisco and Oxford, Westview Press

Fransson, Niklas and Gärling, Tommy 1999 'Environmental Concern: Conceptual Definitions, Measurement Methods, and Research Findings', *Journal of Environmental Psychology* 19, 369–382

Freyer, Hans 1955 *Theorie des gegenwärtigen Zeitalters.* Stuttgart, DVA

Freyer, Hans 1965 'Der Ernst des Fortschritts', in *Technik im technischen Zeitalter,* ed. Freyer, Hans et al. Düsseldorf, Schilling

Gates, Susan 1995 *Außer Kontrolle.* Aarau, Aare (Engl. orig.: *Dragline* 1991)

Geden, Oliver 1999 *Rechte Ökologie. Umweltschutz zwischen Emanzipation und Faschismus.* 2nd edition Berlin, Elefanten Press

Gelberg, Hans-Joachim, ed. 1988 *Die Erde ist mein Haus. 8. Jahrbuch der Kinderliteratur.* Weinheim, Beltz

Gelberg, Hans-Joachim, ed. 1993 *Was für ein Glück. 9. Jahrbuch der Kinderliteratur.* Weinheim, Beltz

George, Edith 1987 'Der Natur das Fell gerben? – Bemerkungen zur Darstellung der Mensch–Natur–Beziehung im Kinderbuch der DDR' in *Resultate (4). Theoretische Schriften zur Kinder- und Jugendliteratur. Die Phantasie und ihre Wunder.* Berlin (East), Der Kinderbuchverlag, 49–61

George, Jean Craighead 1973 *Rotkehlchen hat gesungen.* Aarau, Sauerländer (Am. orig.: *Who Really killed Cock Robin* 1971)

George, Jean Craighead 1974 *Julie von den Wölfen.* Aarau, Sauerländer (Munich, Deutscher Taschenbuch Verlag, 1979 [=dtv junior, Vol. 7351]) (Am. orig.: *Julie of the Wolves* 1972)

George, Jean Craighead 1985 *Stimme aus den großen Sümpfen.* Solothurn, Aare (Am. orig. 1983)

George, Jean Craighead 1988 *Der Ruf des weißen Wals.* Solothurn, Aare (Am. orig.: *Water Sky* 1987)

George, Jean Craighead 1993 *Vermißt im Gumbo Limbo. Ein Umwelt-Krimi.* Aarau, Aare/Sauerländer (Am. orig.: *The Missing 'Gator of Gumbo Limbo* 1992)

George, Jean Craighead 1997 *Julie.* Aarau, Sauerländer (Am. orig.: *Julie* 1994)

Gibowski, Wolfgang 1999 'Social Change and the Electorate: An Analysis of the 1998 Bundestagswahl', *German Politics* 8, 2, 10–32

Gilsenbach, Reimar 1994 *Trostlied für Mäuse. Grüne Lieder, gesungen in Zorn und Liebe.* Eberswalde, published privately

Glotfelty, Cheryll and Fromm, Harold 1996 *The Ecocriticism Reader. Landmarks in Literary Ecology.* Athens, Georgia and London, University of Georgia Press

Göbel, Eberhard and Guthke, Beate 1979 'Die tägliche Revolution in den Oasen der Freiheit – eine gesellschaftliche Alternative?', *Das Argument* 21, 865–870

Goodbody, Axel 1997 'Literature on the environment in the GDR. Ecological activism and the aesthetics of literary protest', in *Retrospect and Review: Aspects of the Literature of the GDR, 1976–1990,* ed. Atkins, Robert and Kane, Martin (*German Monitor* 40). Amsterdam and Atlanta, Rodopi, 238–260

Goodbody, Axel 1999 'From Raabe to Amery: German Literature in Ecocritical Perspective', in *From Classical Shades to Vickers Victorious: Shifting Perspectives in British German Studies,* ed. Giles, Steve and Graves, Peter. Bern, etc., Peter Lang, 77–96

Goodbody, Axel, ed. 1998 *Literatur und Ökologie* (Amsterdamer Beiträge zur neueren Germanistik 43). Amsterdam and Atlanta, Rodopi

Großklaus, Götz and Oldemeyer, Ernst, ed. 1983 *Natur als Gegenwelt. Beiträge zur Kulturgeschichte der Natur*. Karlsruhe, von Loeper

Gruhl, Herbert 1975 *Ein Planet wird geplündert. Die Schreckensbilanz unserer Politik*. Frankfurt am Main, Fischer

Grüne Charta von der Mainau 1961 [Pamphlet]

Guggenberger, Bernd 1975 *Wohin treibt die Protestbewegung? Junge Rebellen zwischen Subkultur und Parteikommunismus*. Freiburg, Herder

Guggenmos, Josef 1990 *Oh, Verzeihung, sagte die Ameise*. Weinheim, Beltz

Haan, Gerhard de and Udo Kuckartz 1996 *Umweltbewußtsein. Denken und Handeln in Umweltkrisen*. Opladen, Westdeutscher Verlag

Hanuschek, Sven 1998 'Carl Amery' in *Kindlers Neues Literatur Lexikon*, ed. Radler, Rudolf, vol. 21. *Supplement A–K*. Munich, 30f.

Harré, Rom, Brockmeier, Jens and Mühlhäusler, Peter 1999 *Greenspeak. A Study of Environmental Discourse*. Thousand Oaks, etc., Sage

Hasenclever, Wolf-Dieter 1992 'Die Grünen im Landtag von Baden-Württemberg. Bilanz nach zwei Jahren Parlamentspraxis' in *Die Grünen. Regierungspartner von morgen?*, ed. Mettke, Jörg R. Reinbek, Rowohlt, 101–119

Heidegger, Martin 1990 'Die Frage nach der Technik', in Heidegger, Martin, *Vorträge und Aufsätze*. 6th edition, Pfullingen, Günther Neske

Hentig, Hartmut von 1978 'Vorwort zur deutschen Ausgabe' in Ariès 1978, 7–44

Herf, Jeffrey 1984 *Reactionary Modernism. Technology, Culture and Politics in Weimar and the Third Reich*. Cambridge, Cambridge University Press

Hermand, Jost 1991 *Grüne Utopien in Deutschland. Zur Geschichte des ökologischen Bewußtseins*. Frankfurt am Main, Fischer

Hermand, Jost and Steakley, James, ed. 1996 *Heimat, Nation, Fatherland: The German Sense of Belonging*. New York, Peter Lang

Herwig, Holger H. 1999 '*Geopolitik*: Haushofer, Hitler and Lebensraum', *Journal of Strategic Studies* 22, 2–3, 218–241

Herzberg, Annegret, ed. 1991 *Staatsmorast. 21 Autoren zur Umwelt*. Lübeck, a & i weißenhorn

Herzinger, Richard and Stein, Hannes 1995 *Endzeit-Propheten oder Die Offensive der Antiwestler. Fundamentalismus Antiamerikanismus und Neue Rechte*. Reinbek, Rowohlt

Hesse, Karen 1997 *Phoenix Rising*. Munich, Deutscher Taschenbuch Verlag (Am. orig. 1994)

Heyne, Isolde 1991 *Wenn die Nachtigall verstummt*. Würzburg, Arena

Himmelheber, Max, Dahl, Jürgen and Löw ,Reinhard 1984/85 'Mitteilung der Redaktion', *Scheidewege* 14, 389

Hines, Jody M., Hungerford, Harold R. and Tomera, Andrey N. 1987 'Analysis and Synthesis of Research on Environmental Behaviour. A Meta-Analysis', *Journal of Environmental Education* 18, 1–8

Hitler, Adolf 1935 *Mein Kampf*. 164th printing, Munich, Zentralverlag der NSDAP Franz Eher Nachfolger

Hoffmann, Jürgen 1998 *Die doppelte Vereinigung. Vorgeschichte, Verlauf und Auswirkungen des Zusammenschlusses von Grünen und Bündnis 90*. Opladen, Leske und Budrich

Horkheimer, Max and Adorno, Theodor Wiesengrund 1973 *Dialectic of Enlightenment*. London, Allen Lane

Hormann, Hanna 1995 'Natur – Umwelt – Ökologie in der Kinder- und Jugendliteratur der DDR' in Dahrendorf 1995, 108–114

Huber, Josef 1980 *Wer soll das alles ändern? Die Alternativen der Alternativbewegung*. Berlin, Rotbuch

Illich, Ivan D. 1973 *Tools for Conviviality*. London, Calder and Boyars

Inglehart, Ronald 1971 'The Silent Revolution in Europe. Intergenerational Change in Post-Industrial Societies', *American Political Science Review* 65, 991–1017

Inglehart, Ronald 1977 *The Silent Revolution. Changing Values and Political Styles among Western Publics*. Princeton, Princeton University Press

Jacobs, Michael, ed. 1997 *Greening the Millennium? The New Politics of the Environment*. Oxford, Blackwell

Jacobs, Thomas 1992 'Visuelle Traditionen des Bergfilms. Von Fidus zu Friedrich oder Das Ende bürgerlicher Fluchtbewegungen im Faschismus' in Amann, Gabel and Keiper 1992, 28–38

Jahn, Detlef 1997 'Green Politics and Parties in Germany' in Jacobs 1997, 174–182

Jahn, Thomas and Wehling, Peter 1990 *Ökologie von rechts. Nationalismus und Umweltschutz bei der Neuen Rechten und den 'Republikanern'*. Frankfurt am Main, Campus

Jamison, Andrew, Eyerman, Ron, Cramer, Jacqueline et al. 1990 *The Making of the New Environmental Consciousness. A Comparative Study of the Environmental Movements in Sweden, Denmark and the Netherlands*. Edinburgh, Edinburgh University Press

Jänicke, Martin 1982 'Parlamentarische Entwarnungseffekte? Zur Ortsbestimmung der Alternativbewegung' in *Die Grünen. Regierungspartner von morgen?*, ed. Mettke, Jörg R. Reinbek, Rowohlt, 69–81

Jefferies, Matthew 1997 'Heimatschutz: Environmental Activism in Wilhelmine Germany' in Riordan 1997a, 42–54

Jesse, Eckhard 1989 'Der politische Prozeß in Deutschland' in *Deutschland-Handbuch. Eine doppelte Bilanz 1949–1989*, ed. Weidenfeld, Werner and Zimmermann, Hartmut. Munich, Hanser, 488–508

Jesse, Eckhard 1992 'Die Entwicklung des Parteiensystems und der Parteien in der Bundesrepublik Deutschland' in *Parteien in Deutschland*, ed. Hübner, Emil and Oberreuter, Heinrich. Munich, Bayerische Landeszentrale für politische Bildungsarbeit, 11–87

Jonas, Hans 1979 *Das Prinzip Verantwortung. Versuch einer Ethik für die technologische Zivilisation*. Frankfurt am Main, Insel

Jowell, Roger, Curtice, John, Park, Alison and Thomson, Katarina 1999 *British Social Attitudes. The 16th Report. Who Shares New Labours values?* Aldershot, Brookfield (USA), Singapore and Sydney, Ashgate

Jünger, Friedrich Georg 1946 *Die Perfektion der Technik*. Frankfurt, Klostermann

Jünger, Friedrich Georg and Himmelheber, Max 1971/72 'Zur Einführung', *Scheidewege* 1, 1–9

Jünger, Friedrich Georg and Himmelheber, Max 1972/73 'Zum zweiten Jahrgang', *Scheidewege* 2, 1–7

Jungk, Robert 1994 *Trotzdem. Ein Leben für die Zukunft*. Munich, Droemer Knaur

Kahlenberg, Friedrich P., Koch, Gertrud, Kreimeier, Klaus and Schlüpmann, Heide 1985 '"Deswegen waren unsere Muttis so sympathische Hühner" (Edgar Reitz). Diskussion zu *Heimat* mit Friedrich P. Kahlenberg, Gertrud Koch, Klaus Kreimeier, Heide Schlüpmann', *Frauen und Film* 38, 96–106

Kaiser, Florian G., Wölfing, Sybille and Fuhrer, Urs 1999 'Environmental Attitude and Ecological Behaviour', *Journal of Environmental Psychology* 19, 1–19

Kaiser, Reinhard, ed. 1981 *Global 2000: Der Bericht an den Präsidenten*. Frankfurt am Main, Zweitausendeins

Kaschuba, Wolfgang et al. 1989 *Der deutsche Heimatfilm, Bildwelten und Weltbilder. Bilder, Texte, Analysen zu 70 Jahren deutscher Filmgeschichte*. Tübingen, Tübinger Vereinigung für Volkskunde

Kaufmann, Eva 1984 'Interview mit Irmtraud Morgner', *Weimarer Beiträge* 30, 1494–1514

Kegel, Bernhard 1997 *Wenzels Pilz. Roman*. Zürich, Amann

Kelly, Petra 1982 'Das System ist bankrott – eine neue Kraft muß her' in *Prinzip Leben*, ed. Kelly, Petra and Leinen, Jo. Berlin, Olle & Wolter

Kelly, Petra 1984 *Fighting for Hope*. London, Chatto & Windus, Hogarth

Kerner, Charlotte 1989 *Geboren 1999. Eine Zukunftsgeschichte*. Weinheim, Beltz

Kerridge, Richard 1999 'BSE Stories', *KeyWords* 2 (Spring), 111–121

Kiermeier-Debre, Joseph, ed. 1996 *Carl Amery. '…ahnen, wie das alles gemeint war'. Ausstellung eines Werkes*. Munich and Leipzig, List

Kirsten, Wulf 1986 *die erde bei meißen.* Leipzig, Reclam

Klages, Helmut 1993 'Wertewandel als Zukunftsperspektive' in *Repräsentative Demokratie und politische Partizipation*, ed. Hirscher, Gerhard. Munich, Hanns-Seidel-Stiftung, 41–58

Klages, Ludwig 1969 *Der Geist als Widersacher der Seele.* Bonn, Bouvier

Klein, Ansgar, Legrand, Hans-Josef and Leif, Thomas, ed. 1999 *Neue Soziale Bewegungen. Impulse, Bilanzen und Perspektiven.* Opladen, Westdeutscher Verlag

Kleinert, Hubert 1992 *Vom Protest zur Regierungspartei. Die Geschichte der Grünen*, Frankfurt am Main, Dietz

Kliewer, Heinz-Jürgen 1995 'Ein Schmetterling ist ein Schmetterling oder Gibt es eine Naturlyrik für Kinder?' in Nassen 1995, 163–175

Knabe, Wilhelm 1995 'Westparteien und DDR–Opposition. Der Einfluß der westdeutschen Parteien in den achtziger Jahren auf unabhängige politische Bestrebungen in der ehemaligen DDR' in *Materialien*, Enquete–Kommission. Vol. VII/2, Baden–Baden, Nomos, 1110–1201

Knapp, Udo 1999 'Die Grünen sind in ihrer Generation eingebunkert', *Deutsches Allgemeines Sonntagsblatt*, 12 March

Koch, Gertrud et al. 1989 'Die fünfziger Jahre: Heide und Silberwald' in Kaschuba et al., ed. 1989, 69–95

Koch, Jurij 1988 *Augenoperation.* Berlin (East), Neues Leben

Koepnick, Lutz P. 1993 'Colonial Forestry: Sylvan Politics in Werner Herzog's *Aguirre* and *Fitzcarraldo*', *New German Critique* 60, 133–59

Kolinsky, Eva, ed. 1989 *The Greens in West Germany: Organisation and Policy Making.* Oxford, Berg

Kracauer, Siegfried 1947 *From Caligari to Hitler: A Psychological History of German Film.* Princeton, Princeton University Press

Kriesi, Hanspeter, Koopmans, Ruud, Duyvendak, Jan Willem and Guigni, Marco G. 1995 *New Social Movements in Western Europe. A Comparative Analysis.* Minneapolis, University of Minneapolis Press

Krönig, Jürgen 2001a 'Hochinfektiös: Der Rinderwahn und die Medien', *EPD Medien* 14, 21 February

Krönig, Jürgen 2001b 'BSE und die Risikogesellschaft', *Berliner Republik* 2, May

Krüger, Christian 1996 'Greenpeace. Politik der symbolischen Konfrontation', in *Forschungsjournal Neue Soziale Bewegungen* 9, 4, 39–47

Kühnel, Wolfgang and Sallmon-Metzner, Carola 1991 'Grüne Partei und Grüne Liga. Der geordnete Aufbruch der ostdeutschen Ökologiebewegung' in *Von der Illegalität ins Parlament. Werdegang und Konzept der Bürgerbewegungen*, ed. Müller-Enbergs, Helmut, Schulz, Marianne and Wielgohs, Jan. Berlin, LinksDruck Verlag, 166–220

Kunert, Günter 1969 *Erinnerung an einen Planeten.* Berlin, Aufbau

Lauermann, Dietmar and Dahl, Jürgen 1989/90 'Max Himmelheber zum Fünfundachtzigsten', *Scheidewege* 19, 1ff.

Lehmann, Jürgen 1999 *Befunde empirischer Forschung zu Umweltbildung und Umweltbewußtsein.* Opladen, Leske und Budrich

Leiner, Friedrich 1981 'Für den Unterrichtspraxis. Der Untergang der Stadt Passau von Carl Amery', *Blätter für den Deutschlehrer* 25, 4, 114–117

Leiner, Friedrich 1982 'Carl Amery: Der Untergang der Stadt Passau. Science-Fiction Roman', in *Deutsche Romane von Grimmelshausen bis Walser. Interpretationen*, ed. Lehmann, Jakob. Frankfurt am Main, Scriptor

Leopold, Aldo 1949 *Sand County Almanac, and Sketches Here and There.* New York, Oxford University Press

Levenstein, Adolf 1912 *Die Arbeiterfrage.* Munich, Reinhardt

Ley, Hermann 1982 'Über die Schwierigkeit der Wirklichkeitsbewältigung', *Deutsche Zeitschrift für Philosophie* 30, 234–247

Lindenpütz, Dagmar 1995 'Umwelt und Ökologie in der Kinder– und Jugendliteratur der BRD' in Dahrendorf 1995, 102–107

Lindenpütz, Dagmar 1999 *Das Kinderbuch als Medium ökologischer Bildung. Untersuchungen zur Konzeption von Natur und Umwelt in der erzählenden Kinderliteratur seit 1970.* Essen, Die Blaue Eule

Lorenc, Kito 1984 *Wortland.* Leipzig, Reclam

Losik, Reinhard 1981 'Alternative Töne von einem DDR-Lyriker', *Frankfurter Rundschau*, 2 June, 4

Lüdke, Hans-Werner and Dinné, Olaf, ed. 1980 *Die Grünen. Personen – Projekte – Programme.* Stuttgart-Degerloch, Seewald

Luhmann, Hans-Jochen 1992 'Warum hat nicht der Sachverständigenrat für Umweltfragen, sondern der SPIEGEL das Waldsterben entdeckt?' in *Jahrbuch Ökologie 1992*, ed. Altner, Günter, Mettler-Meibom, Barbara et al. Munich, Beck, 292–307

Mallinckrodt, Anita 1987 *The Environmental Dialogue in the GDR: Literature, Church, Party, and Interest Groups in their Socio-Political Context.* Lanham, University Press of America

Maloney, Michael P. and Ward, Michael P. 1973 'An Objective Scale for the Measurement of Ecological Attitudes and Knowledge', *American Psychologist* 28, 583–586

Maren-Grisebach, Manon 1984 *Philosophie der Grünen.* Munich and Vienna, Günter Olzog

Maron, Monika 1981 *Flugasche.* Frankfurt, Fischer

Marquand, J.P. 1937 *The Late George Apley. A Novel in the Form of a Memoir.* Boston, Little, Brown and Co.

Marsh, David 1989 *The Germans. Rich, Bothered and Divided.* London, Century

Marx, Leo 1964 *The Machine in the Garden.* New York, Oxford University Press

Maxeiner, Dirk and Miersch, Michael 1996 *Öko-Optimismus.* Düsseldorf and Munich, Metropolitan

Mayer-Pfannholz, Anton 1930 *Wandern und Sehen.* Munich, Oldenbourg

McKibben, Bill 1990 *The End of Nature.* London, Viking

Meadows, Dennis L. et al. 1972 *Die Grenzen des Wachstums. Bericht des Club of Rome zur Lage der Menschheit.* Stuttgart, Deutsche Verlags-Anstalt

Meisner, Mark S. 1995 'Metaphors of Nature. Old Vinegar in New Bottles?', *Trumpeter* 12, 1 (Winter) 11–18

Merchant, Carolyn 1983 *The Death of Nature: Women, Ecology and the Scientific Revolution.* New York, Harper Collins

Meyer-Abich, Klaus Michael 1990 *Aufstand für die Natur. Von der Umwelt zur Mitwelt.* Munich, Hanser

Miller, Walter M., Jr 1960 *A Canticle for Leibowitz.* Philadelphia, Lippincott

Mills, William J. 1982 'Metaphorical Vision: Changes in Western Attitudes to the Environment', *Annals of the Association of American Geographers* 72, 237–253

Mohler, Armin 1974 *Von rechts gesehen.* Stuttgart, Seewald

Morgner, Irmtraud 1974 *Leben und Abenteuer der Trobadora Beatriz nach Zeugnissen ihrer Spielfrau Laura. Roman in dreizehn Büchern und sieben Intermezzos.* Berlin, Aufbau

Morgner, Irmtraud 1983 *Amanda. Ein Hexenroman.* Berlin and Weimar, Aufbau

Morris-Keitel, Peter and Niedermeier, Michael 2000 *Ökologie und Literatur.* New York, Washington, etc., Peter Lang

Müller, Heiner 1958 'Gedanken über die Schönheit der Landschaft bei einer Fahrt zur Großbaustelle "Schwarze Pumpe"', *Junge Kunst* 11, 62

Müller, Jörg 1973 *Alle Jahre wieder saust der Presslufthammer nieder oder Die Veränderung der Landschaft.* Aarau, Sauerländer

Müller, Jörg 1976 *Hier fällt ein Haus, dort steht ein Kran und ewig droht der Baggerzahn oder Die Veränderung der Stadt.* Aarau, Sauerländer

Müller, Jörg and Steiner, Jörg 1976 *Der Bär, der ein Bär bleiben wollte.* Aarau, Sauerländer

Müller, Jörg and Steiner, Jörg 1977 *Die Kanincheninsel.* Aarau, Sauerländer

Müller, Jörg and Steiner, Jörg 1981 *Die Menschen im Meer.* Aarau, Sauerländer

Müller, Jörg and Steiner, Jörg 1983 *Der Eisblumenwald.* Aarau, Sauerländer

Müller, Jörg and Steiner, Jörg 1989 *Aufstand der Tiere oder Die neuen Stadtmusikanten.* Aarau, Sauerländer

Nash, Roderick F. 1989 *The Rights of Nature. A History of Environmental Ethics.* Madison, Wisconsin, University of Wisconsin Press

Nassen, Ulrich 1993 'Trieb, Instinkt, Politik. Einige Tierdarstellungen und –projektionen in fiktionalen Tiererzählungen für Kinder und Jugendliche 1918–1945', *Beiträge Jugendliteratur und Medien*, ed. Heidtmann, Horst. 4th Supplementary Vol., 95–106

Nassen, Ulrich, ed. 1995 *Naturkind, Landkind, Stadtkind. Literarische Bilderwelten kindlicher Umwelt.* Munich, Fink

Neidhardt, Friedhelm and Rucht, Dieter 1993 'Auf dem Weg in die *Bewegungsgesellschaft.* Über die Stabilisierbarkeit von sozialer Bewegung', *Soziale Welt* 44, 3, 305–326

Noelle-Neumann, Elisabeth and Köcher, Renate, ed. 1997 *Allensbacher Jahrbuch der Demoskopie 1993–1997.* Allensbach am Bodensee, Verlag für Demoskopie

Noelle-Neumann, Elisabeth et al. 1983 *Allensbacher Jahrbuch der Demoskopie 1978–1983.* Munich, Saur

North, Richard 1995 *Life on a Modern Planet. A Manifesto for Progress.* Manchester, Manchester University Press

Nöstlinger, Christine 1990 *Nagle einen Pudding an die Wand!* Hamburg, Oetinger

Nowotny, Joachim 1969 *Der Riese im Paradies.* Berlin (East), Der Kinderbuchverlag

Nowotny, Joachim 1981 *Abschiedsdisco.* Berlin (East), Edition Holz (in Der Kinderbuchverlag)

Nowotny, Joachim 1990 *Adebar und Kunigunde. Eine Erzählung für neun Abende...* Berlin, Der Kinderbuchverlag

Opp, Karl-Dieter 1996 'Aufstieg und Niedergang der Ökologiebewegung in der Bundesrepublik' in Diekmann and Jaeger 1996, 350–379

Papadakis, Elim 1984 *The Green Movement in West Germany.* London and Canberra, Croom Helm

Pappi, Franz Urban 1989 'Die Anhänger der neuen sozialen Bewegungen im Parteiensystem der Bundesrepublik', *Aus Politik und Zeitgeschichte*, B 26, 17–27

Pausewang, Gudrun 1983 *Die letzten Kinder von Schewenborn oder ... sieht so unsere Zukunft aus?* Ravensburg, Maier

Pausewang, Gudrun 1987 *Die Wolke. Jetzt werden wir nicht mehr sagen können, wir hätten von nichts gewußt.* Ravensburg, Maier

Pausewang, Gudrun 1990 *Kreuzweg für die Schöpfung.* Baden-Baden, Signal

Pausewang, Gudrun 1991 *Es ist doch alles grün ... Umweltgeschichten nicht nur für Kinder.* Ravensburg, Maier

Pausewang, Gudrun 1993 'Zukünftige Auferstehung' in Gelberg 1993, 313–314

Pech, Klaus-Ulrich, ed. 1985 *Kinder- und Jugendliteratur vom Biedermeier bis zum Realismus.* Stuttgart, Reclam

Pfenning, Uwe 1992 *Politische Netzwerke, Mitgliederstruktur und Aktivitätstypologie der GRÜNEN Rheinland-Pfalz. Eine Fallstudie aus dem Jahr 1984*, Passau, Passauer Papiere zur Sozialwissenschaft 8, University of Passau

Pfister, Christian, ed. 1996 *Das 1950er Syndrom. Der Weg in die Konsumgesellschaft.* Bern, Haupt

Pflaum, Hans Günther and Prinzler, Hans Helmut 1992 *Film in der Bundesrepublik Deutschland. Der neue deutsche Film von den Anfängen bis zur Gegenwart. Mit einem Exkurs über das Kino der DDR.* Munich, Hanser

Pietraß, Richard 1990 *Weltkind.* Leipzig, Reclam

Pirskawetz, Lia 1985 *Der stille Grund.* Berlin, Neues Leben

Pirskawetz, Lia 1987 'Umweltkunst und –literatur sind unverzichtbar', *Natur und Umwelt* 2, 75–77

Pludra, Benno 1980 *Insel der Schwäne.* Berlin (East), Der Kinderbuchverlag

Porritt, Jonathon 1997 'Environmental Politics: The Old and the New' in Jacobs 1997, 47–73

Preisendörfer, Peter 1998 *Umweltbewußtsein in Deutschland 1998. Ergebnisse einer repräsentativen Bevölkerungsumfrage.* Berlin, Federal Ministry for the Environment, Protection of Nature and Nuclear Safety

Preisendörfer, Peter and Franzen, Axel 1996 'Der schöne Schein des Umweltbewußtseins. Zu den Ursachen und Konsequenzen von Umwelteinstellungen in der Bevölkerung' in Diekmann and Jaeger 1996, 219–244

Rabisch, Birgit 1992 *Duplik Jonas 7.* Recklinghausen, Bitter

Rapp, Christian 1997 *Höhenrausch: Der deutsche Bergfilm.* Vienna, Sonderzahl

Raschke, Joachim 1993 *Die Grünen. Wie sie wurden, was sie sind.* Cologne, Bund

Raschke, Joachim 1995 'Diskurs- und andere grüne Schwächen' in *Standortbestimmung bündnisgrüner Politik. Der Strategiekongreß von Bündnis90/ Die Grünen vom 30.9. bis 1.10.1995 in Bonn.* Bonn, Bündnis 90/ Die Grünen, 17–19

Raschke, Joachim 1998 'Machtwechsel und soziale Bewegungen', *Forschungsjournal Neue Soziale Bewegungen,* 11, 1, 25–47

Raschke, Joachim 1999 'Machtwechsel und soziale Bewegungen' in Klein et al. 1999, 64–88

Rentschler, Eric 1986 'The Politics of Vision: Herzog's *Heart of Glass*' in *The Films of Werner Herzog. Between Mirage and History,* ed. Corrigan, Timothy. New York and London, Methuen, 159–81

Rentschler, Eric 1996 'A Legend for Modern Times: *The Blue Light* (1932)' in Eric Rentschler, *The Ministry of Illusion. Nazi Cinema and its Afterlife.* Cambridge, Mass. and London, Harvard University Press, 27–51

Riefenstahl, Leni 1992 *The Sieve of Time: The Memoirs of Leni Riefenstahl.* London, Quartet

Rink, Dieter, in collaboration with Gerber, Saskia 2000 'Institutionalization in lieu of Mobilization. The environmental movement in eastern Germany' in *Social Movements in Central Europe,* ed. Flam, Helena. (forthcoming)

Riordan, Colin 1997a 'Green Ideas in Germany: A Historical Survey' in Riordan 1997b, 3–41

Riordan, Colin, ed. 1997b *Green Thought in German Culture. Historical and Contemporary Perspectives.* Cardiff, University of Wales Press

Riordan, Colin 1999 '"Der Weg in die Zukunft": Uwe Timm and the Problem of Political Ecology' in *Uwe Timm,* ed. Basker, David. Cardiff, University of Wales Press, 66–81

Rippey, Ted, Sundell, Melissa and Towney, Suzanne 1996 '"Ein wunderbares Heute": The Evolution and Functionalization of *"Heimat"* in West German *Heimat* Films of the 1950s' in Hermand and Steakley, ed., 137–59

Rohkrämer, Thomas 1999a 'Antimodernism, Reactionary Modernism and National Socialism. Technocratic Tendencies in Germany 1890–1945', *Contemporary European History* 8/1, 29–50

Rohkrämer, Thomas 1999b *Eine andere Moderne? Zivilisationskritik, Natur und Technik in Deutschland 1880–1933.* Paderborn, Schöningh

Rollins, William 1996 '*Heimat,* Modernity, and Nation in the Early Heimatschutz Movement' in Hermand and Steakley, ed., 87–112

Rootes, Chris 1995 'Environmental consciousness, institutional structures and political competition in the formation and development of Green parties' in *The Green Challenge. The Development of Green Parties in Europe,* ed. Richardson, Dick and Rootes, Chris. London and New York, Routledge, 232–252

Rootes, Chris 1997 'Environmental Movements and Green Parties in Western and Eastern Europe' in *The International Handbook of Environmental Sociology,* ed. Redclift, Michael and Woodgate, Graham. Cheltenham and Northampton, MA, Edward Elgar, 319–347

Rootes, Chris 1999a 'The Transformation of Environmental Activism: activists, organisations and policy-making', *Innovation* 12, 153–175

Rootes, Chris, ed. 1999b *Environmental Protest in Seven European Union States.* Supplementary Report for EC Project No. ENV4–CT97–0514

Rootes, Chris, ed. 1999c *Environmental Movements: Local, National and Global.* Special issue of *Environmental Politics,* 8, 1

Rootes, Chris and Miller, Alexander J. 2000 *The British environmental movement: organisational field and network of organisations*. Paper presented at the ECPR Joint Workshops, Copenhagen, 14–19 April, 2000

Rost, Helga, ed. 1991 *... denn die Natur ist nicht der Menschen Schemel*. Leipzig, Reclam

Roth, Roland 1994 *Demokratie von Unten. Neue Soziale Bewegungen auf dem Wege zur Politischen Institution*. Cologne, Bund

Roth, Roland 1998 'Neue soziale Bewegungen und liberale Demokratie', *Forschungsjournal Neue Soziale Bewegungen*, 11, 1, 48–62

Roth, Roland 1999a 'Ein politischer GAU für die neuen sozialen Bewegungen – Zwischenbilanz nach einem Jahr rot-grüner Bundesregierung', *Forschungsjournal Neue soziale Bewegungen* 12, 10–21

Roth, Roland 1999b 'Neue soziale Bewegungen und liberale Demokratie. Herausforderungen, Innovationen und paradoxe Konsequenzen' in Klein et al., ed. 1999, 47–63

Roth, Roland, and Rucht, Dieter, ed. 1991 *Neue soziale Bewegungen in der Bundesrepublik*. 2nd edition, Frankfurt am Main, Campus

Rowe, J. Stan 1989 'What on Earth is Environment?', *Trumpeter* 6, 4, 123–126

Rucht, Dieter 1987 'Zum Verhältnis von politischen Parteien und sozialen Bewegungen', *Journal für Sozialforschung* 27, 297–313

Rucht, Dieter 1991 'Von der Bewegung zur Institution? Organisationsstrukturen der Ökologiebewegung' in *Neue soziale Bewegungen in der Bundesrepublik Deutschland*, ed. Roth, Roland and Rucht, Dieter. Bonn, Bundeszentrale für politische Bildung, 334–358

Rucht, Dieter 1994 *Modernisierung und neue soziale Bewegungen. Deutschland, Frankreich und USA im Vergleich*. Frankfurt am Main, Campus

Rucht, Dieter 1996 'Wirkungen von Umweltbewegungen: Von den Schwierigkeiten einer Bilanz', *Forschungsjournal NSB*, 9, 4, 15–27

Rucht, Dieter 1999a 'Gesellschaft als Projekt – Projekte in der Gesellschaft. Zur Rolle sozialer Bewegungen' in Klein et al. 1999, 15–27

Rucht, Dieter 1999b 'The Impact of Environmental Movements in Western Societies' in *How Social Movements Matter,* ed. Giugni, Marco, McAdam, Doug and Tilly, Charles. Minneapolis and London, University of Minnesota Press, 204–224

Rucht, Dieter and Roose, Jochen 1999 'The German Environmental Movement at a Crossroads?' in Rootes 1999b, 59–80

Rucht, Dieter and Roose, Jochen 2000a 'Neither Decline nor Sclerosis: The Organisational Structure of the German Environmental Movement', paper presented at the ECPR Joint Workshops, Copenhagen, 14–19 April, 2000

Rucht, Dieter and Roose, Jochen 2000b *Environmental Protest in Germany: A Tale of Stability*. Unpublished manuscript

Rucht, Dieter, Blattert, Barbara and Rink, Dieter 1997 *Soziale Bewegungen auf dem Weg zur Institutionalisierung. Zum Strukturwandel 'alternativer' Gruppen in beiden Teilen Deutschlands*. Frankfurt am Main and New York, Campus

Runge, Gabriele, ed. 1991 *Über Gudrun Pausewang*. Ravensburg, Maier

Sachverständigenrat für Umweltfragen 2000 *Umweltgutachten 2000. Schritte ins nächste Jahrtausend*. Internet: http://www.umweltrat.de/gut00kf1.htm (08/05/00)

Schahn, Joachim 1995 *Die Erfassung und Veränderung des Umweltbewußtseins. Eine Untersuchung zu verschiedenen Aspekten des Umweltbewußtseins und zur Einführung der Wertstofftrennung beim Hausmüll in zwei süddeutschen Kommunen*. Frankfurt am Main, Peter Lang

Schama, Simon 1995 *Landscape and Memory*. London, Harper Collins

Schelsky, Helmut 1961 *Der Mensch in der wissenschaftlichen Zivilisation*. Köln, Westdeutscher Verlag

Scherf, Dagmar 1991 *Geburtstagsbäume für alle. Lebensraum Obstwiese*. Berlin, Klopp

Schildt, Axel 1995 *Moderne Zeiten. Freizeit, Massenmedien und 'Zeitgeist' in der Bundesrepublik der 50er Jahre*. Hamburg, Christians

Schildt, Axel 1999 *Ankunft im Westen. Ein Essay zur Erfolgsgeschichte der Bundesrepublik*. Frankfurt am Main, Fischer

Schiller, Friedrich von 1962 'Ueber naïve und sentimentalische Dichtung' in *Schillers Werke. Nationalausgabe*, Vol. 20, Weimar, Böhlau

Schindler, Regine and Heyduck-Huth, Hilde 1982 *Deine Schöpfung – meine Welt.* Lahr, Kaufmann

Schmidt, Josef 1990 *Parlament und Bewegung. Baden-Württembergs Grüne und die Anti-AKW-Bewegung seit Tschernobyl.* Hannover, Umwelt- und Politik-Verlag

Schmitt-Beck, Rüdiger 1992 'Wertewandel' in *Lexikon der Politik*, ed. Nohlen, Dieter. Vol. 3, *Die westlichen Länder.* Munich, Beck, 527–533

Schneider, Michael 1984 *Nur tote Fische schwimmen mit dem Strom. Essays, Aphorismen, Polemiken.* Cologne, Kiepenheuer and Witsch

Schneider, Wolfgang 1995 'Die Erwachsenen bringen die Kinder um. Umweltutopien in der dramatischen Kinderliteratur' in Nassen 1995, 153–161

Scholes, Katherine and Buchholz, Quint 1990 *Sam's Wal.* Ravensburg, Maier (Austral. orig. 1985)

Schönfelder, Stefan 1999 'Das Verhältnis von Umweltbewegung und Bündnis 90/ Die Grünen in den sechs östlichen Bundesländern am Beispiel der Grünen Liga' in *Bündnis 90/ Die Grünen in den neuen Bundesländern.* Berlin and Potsdam, Heinrich-Böll-Stiftung, 96–99

Schulin, Ernst 1979 *Traditionskritik und Rekonstruktionsversuch: Studien zur Entwicklung von Geschichtswissenschaft und historischem Denken.* Göttingen, Vandenhoeck

Schumacher, Ernst Friedrich 1973 *Small is Beautiful. A Study of Economics as if People Really Mattered.* London, Blond and Briggs

Schwab, Günther 1958 *Der Tanz mit dem Teufel. Ein abenteuerliches Interview.* Hamlyn, Adolf Sponholtz

Schwarzer, Anneliese 1993 *Grünberg lebt.* Wuppertal, Hammer

Seidl, Claudius 1987 *Der deutsche Film der fünfziger Jahre.* Munich, Heyne

Semetko, Holli and Schoenbach, Klaus 1999 'Parties, Leaders and Issues in the News', *German Politics* 8, 2, 72–87

Sheldrake, Rupert 1990 *The Rebirth of Nature: The Greening of Science and God.* London, Century

Sieferle, Rolf Peter 1984 *Fortschrittsfeinde? Opposition gegen Technik und Industrie von der Romantik bis zur Gegenwart.* Munich, Beck

Simmel, Johannes Mario 1990 *Im Frühling singt zum letztenmal die Lerche.* Munich, Droemer Knaur

Soper, Kate 1995 *What is Nature?* Oxford, Blackwell

Spada, Hans 1990 'Umweltbewußtsein: Einstellung und Verhalten' in *Ökologische Psychologie. Ein Handbuch in Schlüsselbegriffen*, ed. Kruse, Lenelis, Graumann, Carl-Friedrich and Lantermann, Ernst-Dieter. Munich, Psychologie Verlags Union, 623–631

Spillner, Wolf 1977 *Gänse überm Reiherberg.* Berlin (East), Der Kinderbuchverlag

Spillner, Wolf 1979 *Der Bachstelzenorden. Fünf Erzählungen.* Berlin (East), Der Kinderbuchverlag

Spillner, Wolf 1984 *Wasseramsel. Die Geschichte von Ulla und Winfried.* Berlin (East), Der Kinderbuchverlag

Spring, Thomas 1986 'Carl Amery. Die ökologische Chance', in 'Buchzeit'. *Sender Freies Berlin*, first and third programmes, 4 March

Statistisches Bundesamt, ed. 1999 *Statistisches Jahrbuch 1999.* Berlin

Statistisches Bundesamt, ed. 2000 *Datenreport 1999. Zahlen und Fakten über die Bundesrepublik Deutschland.* In cooperation with Wissenschaftszentrum Berlin für Sozialforschung (WZB) and Zentrum für Umfragen, Methoden und Analysen (ZUMA). Bonn, Bundeszentrale für politische Bildung

Stern, Paul C. 1992 'Psychological Dimensions of Global Environmental Change', *Annual Review of Psychology* 43, 269–302

Stöss, Richard 1984 'Vom Mythos der "neuen sozialen Bewegungen"' in *Politische Willensbildung und Interessenvermittlung*, ed. Falter, Jürgen W. et al. Opladen, Westdeutscher Verlag, 548–559

Stöss, Richard 1987 'Parteien und soziale Bewegungen. Begriffliche Abgrenzung – Volksparteien – Neue soziale Bewegungen – DIE GRÜNEN' in *Neue soziale Bewegungen in der Bundesrepublik Deutschland*, ed. Roth, Roland and Rucht, Dieter. Bonn, Bundeszentrale für politische Bildung

Stringer, Jenny, ed. 1996 *The Oxford Companion to Twentieth-Century Literature in English*. Oxford and New York, Oxford University Press

Tabbert, Reinbert 1995 'Umweltmythen in Kinderbüchern verschiedener Nationen' in Nassen 1995, 135–151

Tanner, Carmen 1999 'Constraints on Environmental Behaviour', *Journal of Environmental Psychology* 19, 145–157

Taylor, Gordon Rattray 1970 *The Doomsday Book*. London, Thames and Hudson

Tebbutt, Susan 1994 *Gudrun Pausewang in Context. Socially Critical 'Jugendliteratur': Gudrun Pausewang and the Search for Utopia*. Frankfurt am Main, Peter Lang

Thorwartl, Walter 1990 *Der Luchsfelsen*. Vienna, Dachs

Töteberg, Michael and Smith, Stephen W. 1999 'Carl Amery' in *Kritisches Lexikon der deutschsprachigen Gegenwartsliteratur*, ed. Arnold, Heinz Ludwig. 61st fascicle

Trenker, Luis 1961 *Berge und Heimat*. Gütersloh, Bertelsmann

Urban, Dieter 1986 'Was ist Umweltbewußtsein? Exploration eines mehrdimensionalen Einstellungskonstruktes', *Zeitschrift für Soziologie* 15, 5, 363–377

van Hüllen, Rudolf 1990 *Ideologie und Machtkampf bei den Grünen. Untersuchung zur programmatischen und innerorganisatorischen Entwicklung einer deutschen 'Bewegungspartei'*. Bonn, Bouvier

Veen, Hans-Joachim 1988 'Die Grünen als Milieupartei' in *Politik, Philosophie, Praxis. Festschrift für Wilhelm Hennis zum 65. Geburtstag*, ed. Maier, Hans et al. Stuttgart, Klett-Cotta, 454–476

Veen, Hans-Joachim and Hoffmann, Jürgen 1992 *Die Grünen zu Beginn der neunziger Jahre. Profil und Defizite einer fast etablierten Partei*. Bonn and Berlin, Bouvier

Veen, Hans-Joachim and Graf, Jutta 1997 *Rückkehr zu traditionellen Werten?* Sankt Augustin, Konrad-Adeuauer-Stiftung

Vogt, Roland 1990 '"Die Linken haben die Grünen besetzt". Interview mit Roland Vogt' in *Die Grünen. 10 bewegte Jahre*, ed. Schroeren, Michael. Vienna, Ueberreuter, 171–179

Volke, Werner 1986 *Josef Hofmiller. Kritiker, Übersetzer, Essayist*. Marbach, Deutsches Literaturarchiv

Voll, Margarete 1991 'Störfälle. Der Streit um die Verleihung des Jugendliteraturpreises' in Runge 1991, 66–71

von Ditfurth, Hoimar 1985 *So laßt uns denn ein Apfelbäumchen pflanzen. Es ist soweit*. Hamburg, Rasch and Röhring

von Weizsäcker, Carl Friedrich 1988 *Bewußtseinswandel*. Munich, Hanser

von Weizsäcker, Ernst Ulrich 1999 *Das Jahrhundert der Umwelt. Vision: Öko-effizient Leben und Arbeiten*. Frankfurt am Main and New York, Campus

Wagner, Friedrich 1964 *Die Wissenschaft und die gefährdete Welt. Eine Wissenssoziologie der Atomphysik*. Munich, Beck

Waldmann, Klaus 1992 *Umweltbewußtsein und ökologische Bildung*, Opladen, Leske und Budrich

Waugh, Evelyn 1938 *Scoop. A Novel About Journalists*. London, Chapman and Hall

Waugh, Evelyn 1942 *Put Out More Flags*. London, Chapman and Hall

Weichold, Jochen 1993 *Regenbogen, Igel, Sonnenblume. Ökologische Bewegungen und Parteien*. Berlin, Dietz

Weidner, Helmut 1995 *25 Years of Modern Environmental Policy in Germany. Treading a Well-Worn Path to the Top of the International Field*. Discussion paper of the Science Center Berlin, FS II 95–301

Weidner, Helmut and Jänicke, Martin 1998 'Vom Aufstieg und Niedergang eines Vorreiters. Eine umweltpolitische Bilanz der Ära Kohl' in *Bilanz der Ära Kohl*, ed. Wewer, Göttrik. Opladen, Leske und Budrich, 201–28

Wellm, Alfred 1974 *Das Pferdemädchen*. Berlin (East), Der Kinderbuchverlag

Wellm, Alfred 1983 *Das Mädchen mit der Katze*. Berlin (East), Der Kinderbuchverlag

Wenke, Martin Konsumstruktur 1993 *Umweltbewußtsein und Umweltpolitik. Eine makroökonomische Analyse des Zusammenhanges in ausgewählten Konsumbereichen*. Berlin, Duncker & Humblot

Wheatley, Nadia 1991 *Eingekreist. Cols Geschichte*. Weinheim, Beltz (Austral. orig.: *The Blooding* 1987)

White, Lynn Jr 1967 'The Historical Roots of our Ecologic Crisis', *Science* 10 March, 1203–7. (Reprinted in Glotfelty and Fromm 1996: 3–14.)

Wild, Reiner, ed. 1990 *Geschichte der deutschen Kinder- und Jugendliteratur*. Stuttgart, Metzler

Winsemius, Dieuwke 1983 *Das Findelkind vom Watt*. Berlin, Klopp (Munich, Deutscher Taschenbuch Verlag, 1986 [=dtv junior, Vol. 70083]) (Dutch orig. 1980)

Wolf, Christa 1963 *Der geteilte Himmel*. Halle, Mitteldeutscher Verlag

Wolf, Christa 1987 *Störfall. Nachrichten eines Tages*. Berlin and Weimar, Aufbau

Wolff, Bernd 1979 *Biberspur*. Berlin (East), Der Kinderbuchverlag

Wolgast, Heinrich 1950 *Das Elend unserer Jugendliteratur. Ein Beitrag zur künstlerischen Erziehung der Jugend*, ed. Arndt-Wolgast, Elisabeth and Flacke, Walter. Worms, Wunderlich, 7th ed.

Wüst, Jürgen 1993 *Konservatismus und Ökologiebewegung. Eine Untersuchung im Spannungsfeld von Partei, Bewegung und Ideologie am Beispiel der Ökologisch-Demokratischen Partei (ÖDP)*. Frankfurt am Main, IKO

Zentralarchiv für Empirische Sozialforschung, ed. 1999 *The International Social Survey Programme ISSP 1985–1995. Data and Documentation*. Cologne, Zentralarchiv für empirische Sozialforschung

Zimmermann, Peter 1984 *Industrieliteratur der DDR: Vom Helden der Arbeit zum Planer und Leiter*. Stuttgart, Metzler

'Zwölf Jahre *Scheidewege*' 1982/83, *Scheidewege* 12, 713–715

INDEX